Aneurysms-Osteoarthritis Syndrome

T0343348

Aneurysms-Osteoarthritis Syndrome

SMAD3 *Gene Mutations*

Denise van der Linde, MD, MSc, PhD
Department of Cardiology, Erasmus University Medical Center
Rotterdam, The Netherlands

Bart L. Loeys, MD, PhD
Center of Medical Genetics, Antwerp University Hospital
Antwerp, Belgium; and the Department of Human Genetics
Radboud University Medical Center Nijmegen, The Netherlands

Jolien W. Roos-Hesselink, MD, PhD
Department of Cardiology, Erasmus University Medical Center
Rotterdam, The Netherlands

AMSTERDAM • BOSTON • HEIDELBERG • LONDON
NEW YORK • OXFORD• PARIS • SAN DIEGO
SAN FRANCISCO • SINGAPORE • SYDNEY • TOKYO

Academic Press is an imprint of Elsevier

Academic Press is an imprint of Elsevier
125 London Wall, London EC2Y 5AS, United Kingdom
525 B Street, Suite 1800, San Diego, CA 92101-4495, United States
50 Hampshire Street, 5th Floor, Cambridge, MA 02139, United States
The Boulevard, Langford Lane, Kidlington, Oxford OX5 1GB, United Kingdom

Copyright © 2017 Elsevier Inc. All rights reserved.

No part of this publication may be reproduced or transmitted in any form or by any means, electronic or mechanical, including photocopying, recording, or any information storage and retrieval system, without permission in writing from the publisher. Details on how to seek permission, further information about the Publisher's permissions policies and our arrangements with organizations such as the Copyright Clearance Center and the Copyright Licensing Agency, can be found at our website: www.elsevier.com/permissions.

This book and the individual contributions contained in it are protected under copyright by the Publisher (other than as may be noted herein).

Notices
Knowledge and best practice in this field are constantly changing. As new research and experience broaden our understanding, changes in research methods, professional practices, or medical treatment may become necessary.

Practitioners and researchers must always rely on their own experience and knowledge in evaluating and using any information, methods, compounds, or experiments described herein. In using such information or methods they should be mindful of their own safety and the safety of others, including parties for whom they have a professional responsibility.

To the fullest extent of the law, neither the Publisher nor the authors, contributors, or editors, assume any liability for any injury and/or damage to persons or property as a matter of products liability, negligence or otherwise, or from any use or operation of any methods, products, instructions, or ideas contained in the material herein.

British Library Cataloguing-in-Publication Data
A catalogue record for this book is available from the British Library

Library of Congress Cataloging-in-Publication Data
A catalog record for this book is available from the Library of Congress

ISBN: 978-0-12-802708-0

For information on all Academic Press publications
visit our website at https://www.elsevier.com/

Working together
to grow libraries in
developing countries

www.elsevier.com • www.bookaid.org

Publisher: Mica Haley
Acquisitions Editor: Stacey Masucci
Editorial Project Manager: Sam W. Young
Production Project Manager: Chris Wortley
Designer: Matthew Limbert

Typeset by Thomson Digital

Contents

4C. Ehlers-Danlos Syndrome
B.L. Loeys, MD, PhD

4D. Bicuspid Aortic Valve
A.L. Duijnhouwer, MD, A.E. van den Bosch, MD, PhD

4E.   Turner Syndrome
A.T. van den Hoven, BSc, J.W. Roos-Hesselink, MD, PhD,
J. Timmermans, MD

5.   Cardiovascular Imaging in Aneurysm-Osteoarthritis Syndrome
R.G. Chelu, MD, D. van der Linde, MD, PhD, K. Nieman, MD, PhD

6.   Treatment Options
D. van der Linde, MD, MSc, PhD, J.W. Roos-Hesselink, MD, PhD

6A.   Optimal Cardiovascular Medical Treatment
B.L. Loeys, MD, PhD

List of Contributors

J.A. Bekkers, MD, PhD, Erasmus University Medical Center, Rotterdam, The Netherlands

P.K. Bos, MD, PhD, Erasmus University Medical Centre, Rotterdam, The Netherlands

R.G. Chelu, MD, Erasmus University Medical Centre, Rotterdam, The Netherlands

J. de Backer, MD, PhD, University Hospital Ghent, Ghent, Belgium

A.L. Duijnhouwer, MD, Radboud University Medical Centre, Nijmegen, The Netherlands

B.L. Loeys, MD, PhD, Antwerp University Hospital/University of Antwerp, Antwerp, Belgium; Radboud University Medical Centre, Nijmegen, The Netherlands

K. Nieman, MD, PhD, Erasmus University Medical Centre, Rotterdam, The Netherlands

J.W. Roos-Hesselink, MD, PhD, Erasmus University Medical Centre, Rotterdam, The Netherlands

J. Timmermans, MD, Radboud University Nijmegen Medical Centre, Nijmegen, The Netherlands

I.M.B.H. van de Laar, MD, PhD, Erasmus University Medical Centre, Rotterdam, The Netherlands

A.E. van den Bosch, MD, PhD, Erasmus University Medical Centre, Rotterdam, The Netherlands

A.T. van den Hoven, BSc, Erasmus University Medical Centre, Rotterdam, The Netherlands

D. van der Linde, MD, MSc, PhD, Erasmus University Medical Centre, Rotterdam, The Netherlands

H.J.M. Verhagen, MD, PhD, Erasmus University Medical Center, Rotterdam, The Netherlands

J.M.A. Verhagen, MD, Erasmus University Medical Centre, Rotterdam, The Netherlands

M.W. Wessels, MD, PhD, Erasmus University Medical Centre, Rotterdam, The Netherlands

Preface

In *Aneurysms-Osteoarthritis Syndrome: SMAD3 Gene Mutations*, we seek to provide the reader with a practical approach to the clinical management of patients with Aneurysms-Osteoarthritis syndrome (AOS) and related disorders. With the rapid pace of developments in the field of heritable thoracic aortic diseases, simply keeping up-to-date with the latest research and putting this information in a practical context is a challenging task. With a dedicated team of editors and contributors, *Aneurysms-Osteoarthritis Syndrome: SMAD3 Gene Mutations* is an essential guide for cardiologists, (clinical) geneticists, cardiothoracic and vascular surgeons, orthopedic surgeons, radiologists, neurologists, fellows, and researchers who seek a contemporary update and overview of the entire spectrum of aortic aneurysm syndromes.

Starting from the genetic etiology of this disease, we take you on a journey through the numerous cardiovascular and systemic features that can be encountered. We outline how to diagnose them and how to treat and manage your AOS patients in daily clinical practice. We describe genotype and phenotype correlations and provide insights into the role of TGF-beta signaling in aortic disease. A practical algorithm summarizing screening and follow-up guidelines for the multidisciplinary clinical care for AOS patients is also included.

In the evolving field of heritable thoracic aortic diseases, the heterogeneous genetic causes of syndromes are currently being unraveled. Many overlapping clinical features make it increasingly difficult to label patients with a specific disease or syndrome. Therefore, an important part of this textbook is dedicated to the differential diagnosis of heritable thoracic aortic diseases, with subchapters focused on Marfan, Loeys-Dietz, Ehlers-Danlos, bicuspid aortic valve, and Turner syndromes.

Only key references are included so that readability is not inhibited by overly dense text. This textbook is made visually appealing by the use of color images, informative tables, and algorithm flow charts.

We give special thanks to our contributors for their devoted efforts, without which this text would not have been possible. Furthermore, we wish to acknowledge the great help provided by the editorial staff at Elsevier Publishing.

D. van der Linde
B.L. Loeys
J.W. Roos-Hesselink

Chapter 1

Genetics of Aneurysms-Osteoarthritis Syndrome

I.M.B.H. van de Laar, MD, PhD, B.L. Loeys, MD, PhD

1 GENETICS

In 2011, we investigated four generations of a family of Dutch origin, with 22 individuals presenting arterial aneurysms and dissections and/or skeletal or cutaneous abnormalities. The segregation reflected autosomal dominant inheritance and variable expression (Fig. 1.1; family 1). Aiming to map the disease gene, we performed a genome-wide linkage analysis using 250k SNP arrays and obtained a significant multipoint LOD [logarithms (base 10) of odds] score of 3.6 on chromosome 15q22.33. The 12.8-Mb candidate region that was identified by fine-mapping contained an interesting candidate gene involved in the transforming growth factor-beta (TGF-β) signaling pathway—namely, the *SMAD3* gene. In family 1, a heterozygous *SMAD3* mutation—c.859C > T (p.Arg287ArgTrp)—was found to segregate with the phenotype. To evaluate the frequency of *SMAD3* mutations among individuals with aneurysms, we sequenced all *SMAD3* exons in 99 individuals with thoracic aortic aneurysms and dissections and features similar to those in patients with Marfan syndrome (MFS) but without *FBN1*, *TGFBR1*, and *TGFBR2* mutations. We found heterozygous *SMAD3* mutations in 2 out of 99 cases—c.741–742delAT (p.Thr247ProfsX61) and c.782C > T (p.Thr261Ile) (Fig. 1.1; families 2 and 3) [1].

Later, our *SMAD3* sequence analysis of 393 patients with thoracic aortic aneurysms and dissections (without mutations in the *FBN1*, *TGFBR1*, and *TGFBR2* genes) revealed five additional novel heterozygous *SMAD3* mutations: c.313delG (p.Ala105ProfsX11), c.539_540insC (p.Pro180ThrfxX7), c.788C > T (p.Pro263Leu), c.1045G > C (p.Ala349Pro), and c.1080dupT (p.Glu361X) (Fig. 1.1; families 4–8) [2]. All missense mutations segregated with the phenotype, affected by highly conserved amino acids, were predicted to be pathogenic by four computerized algorithms and absent from controls. Other pathogenic mutations introduced a frameshift or stop codon.

The incidence of *SMAD3* mutations in the thoracic aortic aneurysms and dissections cohort seems rather rare, because it was found in only 1–2% of our

Aneurysms-Osteoarthritis Syndrome. http://dx.doi.org/10.1016/B978-0-12-802708-0.00001-6
Copyright © 2017 Elsevier Inc. All rights reserved.

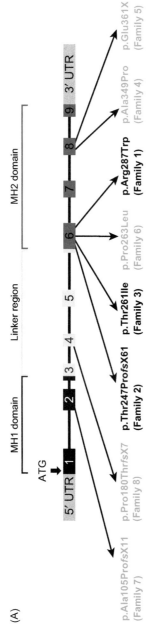

FIGURE 1.1   *SMAD3* **mutations in eight families with Aneurysms-Osteoarthritis syndrome (AOS).** (A) Schematic representation of the *SMAD3* gene. Boxes represent exons 1–9 with the untranslated regions (UTRs). The three main functional domains—MH1, MH2, and the linker region—are indicated. Mutations previously identified in the AOS are depicted in black font, and mutations identified in this study are depicted in blue. (B) Simplified family trees of eight unrelated families with AOS. *Squares* indicate males, and *circles* represent females. A *horizontal line* above the symbol indicates medical examination by one of us. Owing to the lack of space, generation III from family 1 is split into two levels. An *arrow* points to the index patient. The *upper-right blue square* indicates the presence of osteoarthritis, the *lower-right red square* the presence of a thoracic aortic aneurysm, the *lower-left green square* the presence of an aneurysm in any other artery, and the *upper-left yellow square* the presence of arterial tortuosity. *Open symbols* are individuals with a normal or unknown phenotype. Four individuals with *open symbols* (family 1, patient III-10, V-5, V-12; and family 3, patient III-2) had other signs of AOS not indicated in the legend. A *question mark* (?) indicates sudden cardiovascular death, possibly from an arterial rupture or dissection without autopsy. Age of death is displayed below the symbol. The presence (+/−) or absence (−/−) of a *SMAD3* mutation is indicated underneath. *(Reprinted with permission from the article by Van de Laar et al., Phenotypic spectrum of the SMAD3-related aneurysms-osteoarthritis syndrome. J Med Genet 2012;49(1):47–57.)*

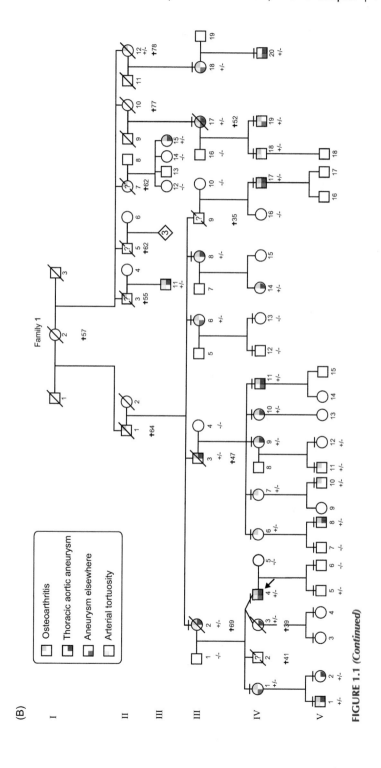

FIGURE 1.1 *(Continued)*

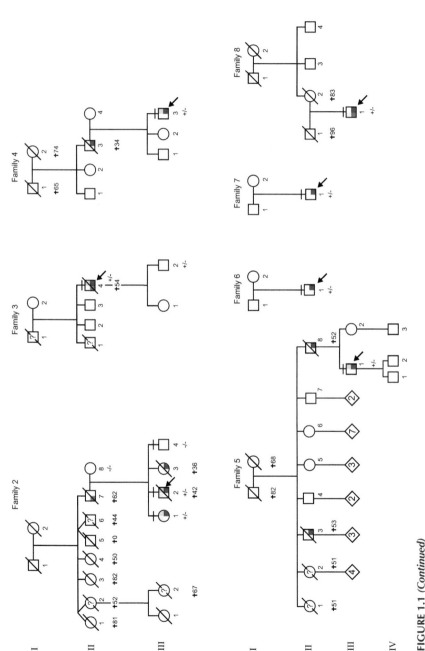

FIGURE 1.1 *(Continued)*

cohort. This rate is comparable with the 2% frequency of mutations in the cohort of nonsyndromic familial thoracic aortic aneurysm and dissection patients reported by Ellen S. Regalado et al. [3]. However, recent studies have revealed a slightly higher incidence (3–4%) of *SMAD3* mutations in a large cohort of both syndromic and nonsyndromic thoracic aortic aneurysm and dissection patients [4].

As of now, 36 different *SMAD3* gene mutations have been published in the literature, but many more unpublished SMAD3 mutations have been identified. Yvonne Hilhorst-Hofstee et al. reported a small interstitial deletion of chromosome 15, leading to disruption of the *SMAD3* gene [5]. The *SMAD3* gene contains three main functional domains—namely, the MH1 and the MH2 domains and the linker region, with mutations occurring throughout the entire 9 exon-containing gene. The most likely effect of these mutations is a loss of function, with TGF-β signals not being propagated via SMAD3. Until now, no clear genotype–phenotype correlation has been established.

2 PATHOPHYSIOLOGY

The *SMAD3* gene encodes the SMAD3 protein, a member of the TGF-β pathway that is crucial for TGF-β signal transmission. We investigated the effect of *SMAD3* mutations on the aortic wall via the histology and immunohistochemistry of aorta fragments obtained during surgery or autopsy. Disorganization of the tunica media with fragmentation and loss of elastic fibers, mucoid medial degeneration, and accumulation of collagen in the media were observed with varying degrees of severity (Fig. 1.2).

We also studied the expression of several members of the TGF-β pathway, including total SMAD3 (nonphosphorylated and phosphorylated forms), phosphorylated SMAD2 (pSMAD2), TGF-β1 and connective tissue growth factor (CTGF), by immunohistochemistry. Despite the loss-of-function nature of the *SMAD3* mutations, the patient-derived aortic tissues showed evidence of increased (rather than decreased) TGF-β signaling, as was observed by the increased labeling intensity of all the studied markers. TGF-β1 expression was present throughout the aneurismal aortic media, whereas the controls only showed substantial expression in the media adjacent to the adventitia layer, which normally shows the highest level of activity (Fig. 1.3).

CTGF immunolabeling showed a markedly increased cytoplasmatic expression in the medial vascular smooth muscle cells (VSMCs) of the cases (Fig. 1.3). This upregulation of both the upstream ligands and the downstream targets of the TGF-β pathway in the thoracic aortic wall of the AOS cases was similar to that of patients with other syndromic and nonsyndromic aneurysms, including MFS, Loeys-Dietz syndrome, arterial tortuosity syndrome, aneurysms associated with the bicuspid aortic valve, and degenerative aneurismal aortic disease [6–8]. This similarity clearly indicates the existence of common (TGF-β-related) pathogenic mechanisms leading to arterial wall disease.

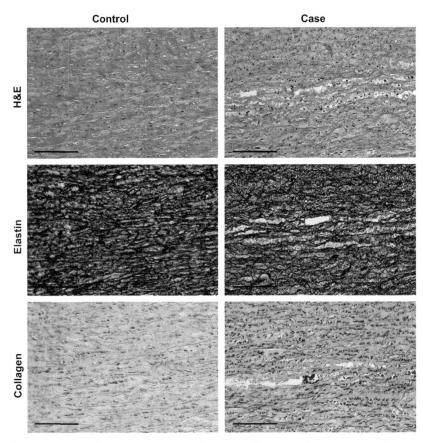

FIGURE 1.2 **Histology of aortic media in Aneurysms-Osteoarthritis syndrome.** Aortic media from a control (donor, left column) and case (right column) with a *SMAD3* mutation resulting in p.Thr247*fs*X61 (III-2, family 2). Scale bars correspond to 100 μm. Hematoxylin–eosin staining displays abnormal architecture of the aortic media and a dissection tear in the case. A Verhoeff-van Gieson staining for elastin *(dark purple fibers)*; note the disarray, fragmentation, and loss of elastic fibers in case versus control. A dissection tear is shown. A Masson's trichrome staining for collagen *(green)* shows intense collagen staining and disruption of the medial architecture in the case. *(Reprinted with permission from the article by Van de Laar et al., Mutations in SMAD3 cause a syndromic form of aortic aneurysms and dissections with early-onset osteoarthritis. Nat Genet 2011;43(2):121–126.)*

Studies in SMAD3 knock-out mice have revealed a phenotype resembling human osteoarthritis, including the abnormal calcification of the synovial joints with osteophytes (knee, vertebral bones, sternum), loss of articular cartilage, intervertebral disc degeneration, and hypertrophic differentiation of articular chondrocytes. These studies confirmed that SMAD3-mediated signals are essential in cartilage maintenance. Later studies have also revealed a vascular phenotype in SMAD3 knock-out mice that is characterized by progressive age-induced aortic root, ascending aorta dilation, aneurysm rupture, and aortic dissection [9].

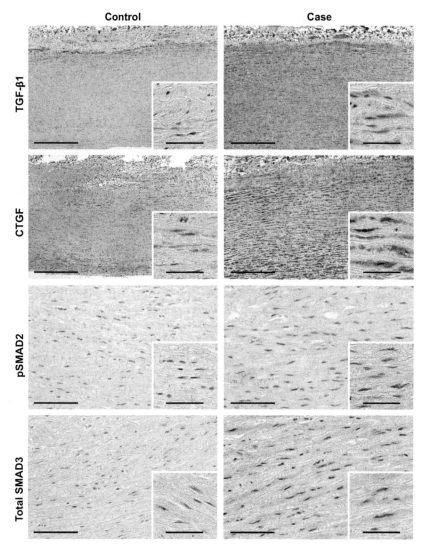

FIGURE 1.3 Immunohistochemistry in Aneurysms-Osteoarthritis syndrome. Aortic wall (from top to bottom, adventitia, media and intima layers) from a control (donor) and a case with a *SMAD3* mutation resulting in p.Arg287Trp (IV-3, family 1). TGF-β1 immunostaining; note the increased TGF-β1 expression through the aortic media, whereas the control only shows marked expression in the outer media adjacent to the adventitia. Connective tissue growth factor (CTGF) immunolabeling is shown. CTGF is a TGF-β-responsive product that normally induces collagen synthesis. Note the increased labeling in the cytoplasm of media cells from the case compared to the donor. Scale bars, 200 μm; inset scale bars, 50 μm. Photomicrographs from the middle section of the aortic media show phosphorylated SMAD2 (pSMAD2) immunolabeling with marked nuclear staining in the case. Total SMAD3 (phosphorylated and nonphosphorylated SMAD3) immunostaining shows increased nuclear and cytoplasmatic labeling in the case as compared to the donor. Scale bars, 100 μm; inset scale bars, 50 μm. *(Reprinted with permission from the article by Van de Laar et al., Mutations in SMAD3 cause a syndromic form of aortic aneurysms and dissections with early-onset osteoarthritis. Nat Genet 2011;43(2):121–126.)*

3 THE TGF-β PATHWAY: THE CANONICAL SIGNALING

The disruption of TGF-β signaling has been implicated in the pathogenesis of many diseases, including aortic aneurysms. This complex pathway has been studied extensively in the past in the context of these diseases, particularly because of its utility as a therapeutic target. TGF-β induces canonical (SMAD-dependent) and noncanonical pathways.

Canonical TGF-β signaling involves different processes, including the ligand binding of TGF-β, receptor recruitment and phosphorylation, SMAD phosphorylation, co-SMAD binding, and the transcription of multiple TGF-β-driven genes.

3.1 Ligand Binding

TGF-β is a ligand of the TGF-β superfamily and is present in three TGF-β isoforms in humans—namely, TGF-β1, TGF-β2, and TGF-β3. TGF-βs are secreted in an inactive form; they are synthesized as propeptide precursors containing a prodomain (also named latency-associated peptide, or LAP) and a mature domain, forming a complex that is called the small latent complex (SLC). By disulfide-bonding the latent TGF-β-binding protein 1 (LTBP-1) to the LAP of the SLC, a larger complex named the large latent complex (LLC) is formed. LTBP-1 possesses domains that interact with matrix molecules, such as microfibrils, thereby targeting the LLC to the extracellular matrix. The interactions between TGF-β1 and LTBP-1 and between LTBP-1 and matrix proteins, respectively, determine the sequestration/release of TGF-β1 within the extracellular matrix, which is an important regulating mechanism of TGF-β signaling.

3.2 Receptor Recruitment and Phosphorylation

The TGF-β receptors have a cysteine-rich extracellular domain, a transmembrane domain, and a cytoplasmic serine/threonine-rich domain. Seven type I receptors or activin-like receptor kinases have been described, including TGFBR1 and five different type II receptors, including TGFBR2. The TGF-β ligand binds to the constitutively active TGF-β type II receptor dimer, which recruits a TGF-β type I receptor dimer to form a complex that facilitates the phosphorylation and the subsequent activation of the type I receptor.

3.3 SMAD Phosphorylation and co-SMAD Binding

SMADs are a well-conserved family of transcriptional factors, and members of the SMAD family can be classified into three groups: (1) receptor-associated SMADs (R-SMADs: namely, SMAD1, SMAD2, SMAD3, SMAD5, and SMAD9), (2) co-operating SMAD (co-SMAD: namely, SMAD4), and (3) inhibitory SMADs (I-SMADs: namely, SMAD6 and SMAD7).

The type I receptor phosphorylates the serine residue of the R-SMAD (eg, SMAD3). Phosphorylation induces a conformational change in the MH2 domain of the R-SMAD and its subsequent dissociation from the receptor complex. The phosphorylated R-SMADs dimerize, and this complex has a high affinity for a co-SMAD (eg, SMAD4) and forms a heterotrimeric complex.

3.4 Transcription

The phosphorylated R-SMAD/co-SMAD complex enters the nucleus of VSMCs, where it binds transcription promotors/cofactors and regulates the expression of multiple TGF-β-driven gene targets, such as matrix proteins (collagen, fibronectin), matrix metalloproteinases (MMPs, and more specifically, MMP2 and MMP9), CTGF, and members of the fibrinolytic system (plasminogen activator inhibitor-1).

4 THE ROLE OF THE TGF-β PATHWAY IN ANEURYSM FORMATION

The pathogenic downstream mechanism of TGF-β is not fully understood. TGF-β signaling has been extensively studied by measuring components of the canonical pathway (SMAD-dependent), such as pSMAD2, and its downstream targets, such as CTGF. However, TGF-β can also induce noncanonical (SMAD-independent) pathways. Studies have reported the involvement of noncanonical SMAD signaling in aneurysm formation through the mitogen-activated kinases, such as the extracellular signal-regulated kinase (ERK)1/2, p38, and the Jun N-terminal kinase (Fig. 1.4) [10].

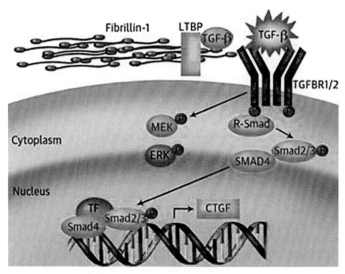

FIGURE 1.4 Schematic display of the transforming growth factor-beta pathway: the canonical signaling. *(Reprinted with permission from the article by Bertoli-Avella AM et al., Mutations in a TGF-β ligand, TGFB3, cause syndromic aortic aneurysms and dissections. J Am Coll Cardiol 2015;65(13):1324–1336.)*

In AOS and other forms of aortic aneurysms, a recurrent signature of enhanced TGF-β signaling is observed at the aortic tissue level despite a loss of function at the molecular level. The precise mechanisms underlying the attenuation of TGF-β signaling remain elusive and a matter of debate. Several mechanisms that could explain this TGF-β paradox have been proposed but need experimental validation. These theories include altered receptor trafficking, impaired autoregulation of TGF-β signaling, alternative signaling cascades, and nonautonomous cellular events.

Several terminal events in the pathogenesis of aneurysms have been proposed. One model that has been experimentally validated is the enhancement of MMP activity leading to proteolysis. TGF-β specifically induces MMP2 and MMP9 expression, the MMPs that are associated with aneurysm formation. Increased MMP expression and activity are seen in many natural and experimentally induced presentations of aneurysm. Other mechanisms include CTGF-mediated epithelial-to-mesenchymal transition and tissue remodeling and IL-6- and MCP-1-mediated inflammation.

REFERENCES

[1] van de Laar IM, Oldenburg RA, Pals G, Roos-Hesselink JW, de Graaf BM, Verhagen JM, et al. Mutations in SMAD3 cause a syndromic form of aortic aneurysms and dissections with early-onset osteoarthritis. Nat Genet 2011;43(2):121–6.

[2] van de Laar IM, van der Linde D, Oei EH, Bos PK, Bessems JH, Bierma-Zeinstra SM, et al. Phenotypic spectrum of the SMAD3-related aneurysms-osteoarthritis syndrome. J Med Genet 2012;49(1):47–57.

[3] Regalado ES, Guo DC, Villamizar C, Avidan N, Gilchrist D, McGillivray B, et al. Exome sequencing identifies SMAD3 mutations as a cause of familial thoracic aortic aneurysm and dissection with intracranial and other arterial aneurysms. Circ Res 2011;109(6):680–6.

[4] Campens L, Callewaert B, Muiño Mosquera L, et al. Gene panel sequencing in heritable thoracic aortic disorders and related entities—results of comprehensive testing in a cohort of 264 patients. Orphanet J Rare Dis 2015;10:9.

[5] Hilhorst-Hofstee Y, Scholte AJHA, Rijlaarsdam MEB, van Haeringen A, Kroft LJ, Reijnierse M, et al. An unanticipated copy number variant ofchromosome 15 disrupting SMAD3 reveals a three-generation family at serious risk for aortic dissection. Clin Genet 2013;83(4):337–44.

[6] Loeys BL, Chen J, Neptune ER, Judge DP, Podowski M, Holm T, et al. A syndrome of altered cardiovascular, craniofacial, neurocognitive and skeletal development caused by mutations in TGFBR1 or TGFBR2. Nat Genet 2005;37(3):275–81.

[7] Wang X, LeMaire SA, Chen L, Shen YH, Gan Y, Bartsch H, et al. Increased collagen deposition and elevated expression of connective tissue growth factor in human thoracic aortic dissection. Circulation 2006;114(Suppl. 1):S200–5.

[8] Gomez D, Al Haj Zen A, Borges LF, Philippe M, Gutierrez PS, Jondeau G, et al. Syndromic and non-syndromic aneurysms of the human ascending aorta share activation of the Smad2 pathway. J Pathol 2009;218(1):131–42.

[9] Ye P, Chen W, Wu J, et al. GM-GSF contributes to aortic aneurysms resulting from SMAD3 deficiency. J Clin Invest 2013;123:2317–31.

[10] Gillis E, Van Laer L, Loeys BL. Genetics of thoracic aortic aneurysm: at the crossroad of transforming growth factor-beta signaling and vascular smooth muscle cell contractility. Circ Res 2013;113(3):327–40.

Chapter 2

Cardiovascular Phenotype of Aneurysms-Osteoarthritis Syndrome

D. van der Linde, MD, MSc, PhD, J.W. Roos-Hesselink, MD, PhD

1 INTRODUCTION

There is no disease more conducive to clinical humility than aneurysm of the aorta.
— William Osler (*Canadian Physician*, 1849–1919)

These words still resonate today, when many patients with ruptured aortic aneurysms die before reaching the hospital and the mortality rate among those who do reach the hospital is still very high. Because aortic aneurysms are often asymptomatic, *"silent killer"* might be a suitable synonym.

Aneurysms-Osteoarthritis syndrome (AOS) is an autosomal dominant inherited syndrome caused by *SMAD3* mutations. The syndrome is characterized by arterial aneurysms and dissections, osteoarthritis, and mild craniofacial features. The cardiovascular traits associated with AOS are often asymptomatic until the development of a catastrophic presentation with an aortic dissection or rupture [1–3]. Thus, key to understanding the syndrome is the knowledge that although the vast majority of patients with AOS exhibit (silent) cardiovascular abnormalities, these issues are not the patients' main reason for initially consulting a physician. In more than half of adult patients, joint-related complaints were the initial symptom for which medical advice was sought. None of the patients suspected having an aneurysm syndrome. As is further discussed in Chapter 3, the other phenotypic features of the disease, such as joint complaints and craniofacial features, in combination with a positive family history for aortic events or sudden death should raise awareness of the possibility of AOS as a diagnosis.

Furthermore, when patients did present with cardiovascular abnormalities, they were often in a late or catastrophic manner. Sudden death from aortic dissections, aortic aneurysms, and severe mitral valve insufficiencies were the most common presentations. In some patients, the diagnosis of Marfan syndrome (MFS) was made at the time of presentation on the basis of the revised Ghent criteria [4].

Aneurysms-Osteoarthritis Syndrome. http://dx.doi.org/10.1016/B978-0-12-802708-0.00002-8
Copyright © 2017 Elsevier Inc. All rights reserved.

11

2 AORTIC ANEURYSMS

The most prominent cardiovascular features associated with AOS are arterial aneurysms and tortuosity. Thoracic aortic aneurysms are the most frequent symptom of AOS, with a reported prevalence between 65% and 80% [1–3,5,6]. In approximately 10% of patients, the abdominal aorta is dilated.

The main site of thoracic aortic aneurysms in AOS patients is the sinus of Valsalva (Fig. 2.1A). The average age at which patients are diagnosed with aneurysms is 39–42 years [1–3,5]. At diagnosis, aortic root diameters range from 36 mm (Z-score 3) to 63 mm (Z-score 13), with a mean diameter of 40 mm. The youngest children with thoracic aortic aneurysms are 15 and 16 years old, with Z-scores of 2.9 and 3.3, respectively.

Annualized progression of aortic dilation in patients with AOS is found to be highest in the sinus of Valsalva, a rate of approximately 2.5 mm per year [7]. Although this estimate is based on a small number of AOS patients and should be confirmed in the future by larger studies with longer follow-up intervals, it has become clear that aortic growth in AOS can be fast and unpredictable. The aortic dilatation progression rate is highly correlated with the initial diameter of the sinus of Valsalva; it is not correlated to age, sex, cholesterol levels, blood pressure, or left ventricular (LV) mass. The annual progression rate seems comparable to or even higher than in patients with MFS or bicuspid aortic valve (BAV) disease [8–16].

3 ARTERIAL ANEURYSMS THROUGHOUT THE BODY

Arterial aneurysms in patients with AOS are not limited to the aorta only but may be present throughout the arterial tree in large- and medium-size vessels. On extensive computed tomography or magnetic resonance imaging, one-third of patients were found to have aneurysms in other thoracic and abdominal arteries, predominantly involving the pulmonary, splenic, iliac, and mesenteric arteries (Fig. 2.1B) [2,3]. Ellen Regalado et al. found iliac artery aneurysms in 5% of AOS patients [5]. One case report described a 32-year-old patient presenting with a painful pulsatile mass in his groin, which was then diagnosed as a 69-mm partially thrombosed aneurysm of the common iliac artery [17]. On rare occasions, arterial aneurysms were encountered in hepatic, renal, gastric, gastroduodenal, internal mammary, and femoral arteries [18]. These features are further described in Chapter 6c.

Imaging of the neck and intracranial arteries identified aneurysms in approximately 40% of patients, predominantly in the vertebral, carotid, basilar, and ophthalmic arteries (Fig. 2.1C). In the cohort studied by Regalado et al., only 10% of patients were found to have intracranial aneurysms. These data suggest that the cerebrovascular circulation as well as the entire aorta and thoracic and abdominal large arteries must be imaged for aneurysms in patients with AOS. Despite the presence of intracranial aneurysms, strokes were not reported in the

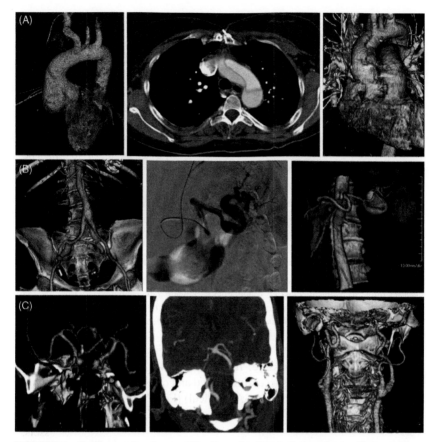

FIGURE 2.1 **Cardiovascular abnormalities throughout the body in patients with Aneurysms-Osteoarthritis syndrome.** (A) Thorax: (left) aneurysm of the aortic root (54 mm) in 31-year-old man; (middle and right) Stanford type A aortic dissection at a maximal aortic diameter of 40 mm in a 50-year-old woman. (B) Abdomen: (left) aortic dissection at a maximal abdominal aortic diameter of 24 mm with dissection flap extending into the left common iliac artery (true lumen in internal iliac artery and false lumen in external iliac artery) and aneurysm in the right common iliac artery (27 mm) and right external iliac artery (16 mm) in 45-year-old woman; (middle and right) tortuosity and aneurysm in left splenic artery (21 mm) in the same 45-year-old woman. (C) Head and neck: (left) two saccular aneurysms in the left and right carotid siphons in a 31-year-old man; (middle) fusiform aneurysm of the top of the basilar artery in a 26-year-old man; (right) tortuosity of the internal carotid artery in a 34-year-old man. *(Reprinted with permission from the article by van der Linde D et al. Aggressive cardiovascular phenotype of aneurysms-osteoarthritis syndrome caused by pathogenic SMAD3 variants. J Am Coll Cardiol 2012;60(5):397–403.)*

Rotterdam cohort [2,3]. However, Regalado et al. reported one patient who died from a subarachnoid hemorrhage [5].

One patient, a 26-year old man, was found to have a dilated pulmonary trunk of 50 mm and a saccular aneurysm of 18×14 mm in a persistent ductus arteriosus (Fig. 2.2) [19]. During catheterization, the pressure in the aneurysm was

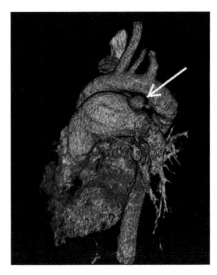

FIGURE 2.2 Saccular aneurysm within a persistent ductus arteriosus and placement of a vascular plug. 3D reconstruction of CT angiography showing a dilated pulmonary trunk (50 mm) and a saccular aneurysm of a persistent ductus arteriosus *(arrow)*. *(Reprinted with permission from the article by van der Linde D et al. Saccular aneurysm within a persistent ductus arteriosus. Lancet 2012;379(9816):e33.)*

75% of systemic arterial pressure. To prevent further enlargement and possible rupture, the aneurysm was successfully filled and closed off from circulation by an Amplatzer Vascular Plug II (AGA Medical) (Fig. 2.3).

4 ARTERIAL TORTUOSITY

Aortic and iliac tortuosity is defined as described by Elliot Chaikof et al. [20]. Visceral arterial tortuosity is defined as a severe (pigtail-like) curve or multiple curves in an artery. Tortuosity of the large- or medium-size arteries was present in the majority of aneurysm-osteoarthritis syndrome patients. In the Rotterdam cohort, aortic tortuosity was found in 38%; tortuosity of other thoracic and the abdominal arteries was also found in 38%; and tortuosity of the carotid and cerebral arteries was found in half of AOS patients [2,3]. Arterial tortuosity was most frequently located in the vertebral, iliac, splenic, carotid, and intracranial arteries (Fig. 2.4). A French cohort of 34 aneurysm-osteoarthritis syndrome patients experienced carotid tortuosity in 72% and tortuosity in other arteries in 11% [6]. In addition, Yvonne Hilhorst-Hofstee et al. described aortic and carotid tortuosity in 60% of patients in a case report of a family of 5 patients [21].

5 DISSECTIONS AND CARDIOVASCULAR MORTALITY

The most feared complication of asymptomatic aortic aneurysms is aortic dissection or rupture (Fig. 2.1A, right panel). Of the initial cohort of patients, approximately one in three patients exhibited an aortic dissection [2]. In about

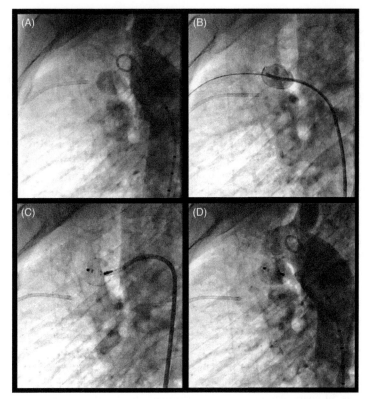

FIGURE 2.3 Angiography images showing different stages of the catheterization procedure. (A) aneurysm of the persistent ductus arteriosus (14 × 18 mm); (B) catheter positioned within the aneurysm; (C) delivery of the vascular plug (16 × 12 mm); (D) closure of the persistent ductus arteriosus with the vascular plug in place. *(Reprinted with permission from the article by van der Linde D et al. Saccular aneurysm within a persistent ductus arteriosus. Lancet 2012;379(9816):e33.)*

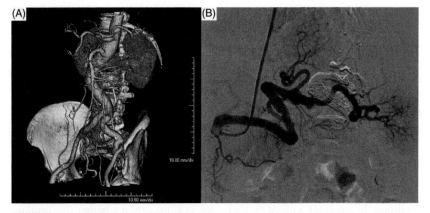

FIGURE 2.4 Arterial tortuosity (A) in the aorta, visceral, and iliac arteries and (B) in the splenic artery. *(Reprinted with permission from the article by van der Linde D et al. Aneurysm-osteoarthritis syndrome with visceral and iliac artery aneurysms. J Vasc Surg 2013;57(1):96–102.)*

15–20% of patients, a dissection was the first cardiovascular manifestation of the disease.

All dissections in the Rotterdam cohort occurred in adulthood at the mean age of 46 ± 10 years; the youngest patient was 34 years of age [1–3]. The average age of onset of dissection in the cohort studied by Regalado et al was 42 years (range 25–54 years). [5] The French cohort by Mélodie Aubart et al. described thoracic aortic dissection in one-fourth of patients at a mean age of 47 years (range 22–68 years) [6]. All together, we can conclude that if left undiscovered, AOS most commonly presents in the fourth decade of life with an aortic dissection.

Stanford type A dissections are more common (approximately 70–80%) than type B dissections. Once a dissection has occurred, be aware that patients could develop a second or recurrent dissection (approximately 10%); two patients were diagnosed with a type A and a type B dissection at different times [2].

The range of aortic root measurements before type A aortic dissection occurred was 40–63 mm, with a mean of 51 mm [2]. Occasionally, aortic dissection occurred in aorta that were only mildly dilated, with evidence of two patients dissecting at diameters of 40 and 45 mm, respectively. Sporadically, patients can present with dissections in other arteries without a prior aneurysm— namely, the proximal left anterior descending coronary, common and internal iliac, and superior mesenteric arteries.

In the initial Rotterdam cohort of 44 patients, 15 patients with AOS died suddenly between the ages of 34 and 69 years [2]. The mean age at the time of death was 54 years. Autopsies were performed on six patients and confirmed a Stanford type A dissection in five patients and a Stanford type B dissection in one patient. In seven patients, no autopsies were performed, but three of them were previously known to have had aortic aneurysms or dissections. Regalado et al. reported one case of intracranial hemorrhage as the cause of death [5].

6 CONGENITAL HEART DEFECTS

Congenital heart defects were found in roughly 10% of AOS patients, a profoundly higher number than expected in the general population, among whom the prevalence is approximately 0.8% [22]. Abnormalities included persistent ductus arteriosus, atrial septal defect, pulmonary valve stenosis, and BAV. One remarkable finding was a saccular aneurysm within a persistent ductus arteriosus (Fig. 2.2) [19].

Recently, a case report by Kristi Fitzgerald et al. described a 14-year-old boy with a novel SMAD3 mutation with hypoplastic left heart syndrome, who developed severe aortic dilatation in the neoaortic root after undergoing a Fontan operation [23]. Thus far, this is the only case reporting hypoplastic left heart syndrome in a patient with AOS; therefore, it is uncertain whether this issue is related to the SMAD3 mutation.

As previously mentioned, evidence from mouse studies suggests that transforming growth factor-beta (TGF-β) signaling is essential in the embryogenesis of the heart and if disrupted might lead to congenital heart defects [24,25]. In many mouse models with disrupted TGF-β signaling activities, congenital heart defects were present [24]. In the future, SMAD3 knockdown mice might help explore the mechanism behind the congenital heart defects associated with AOS.

7 MITRAL VALVE ANOMALIES AND ATRIAL FIBRILLATION

Similar to MFS, mitral valve anomalies were common in AOS patients (50%) in the Rotterdam cohort, with the youngest being 14 years old [1–3]. Abnormalities ranged from just a prolapse or billowing without hemodynamic significance (most common) to severe mitral valve regurgitation requiring valve replacement (rare). Mice studies have shown that TGF-β signaling is essential in the embryogenesis of the heart and valvular pathogenesis [24,25]. Regalado et al. reported mitral valve disease in 17% of patients [5].

In addition, AOS patients seem to be prone to developing atrial fibrillation (AF), because about 24% have had at least one episode by the time of diagnosis in their mid-40s [2]. In about one-third of these patients, AF only occurred in a single episode after cardiothoracic surgery. Basic science has thus far provided some possible explanations as to why disruption of the TGF-β pathway in AOS patients might cause AF. Mice studies have shown that TGF-β1-induced myocardial fibrosis in the atria plays an important role in predisposing individuals to AF [26]. Atrial fibrogenesis in patients with AF occurs in two phases: an early increase but a later loss of responsiveness to TGF-β1 as the fibrosis progresses [27].

8 LEFT VENTRICULAR FUNCTION AND HYPERTROPHY

LV systolic function and mitral inflow patterns were normal in all patients in the initial Rotterdam series [2]. However, mild to moderate LV hypertrophy was diagnosed in one-fifth of patients with a LV posterior wall thickness of 12 mm and a mean LV mass of 296 g [2]. Remarkably, none of the patients with LV hypertrophy had hypertension, aortic coarctation, or aortic stenosis. Currently, we can only hypothesize why AOS patients develop LV hypertrophy. Primary cardiomyopathy is reported in one-quarter of MFS patients showing mainly a reduced LV ejection fraction, but only a small minority (2.9%) experienced an increased LV mass [28]. Mice studies have determined that TGF-β induces proliferation of cardiac fibroblasts and hypertrophic growth of cardiomyocytes [29]. Furthermore, TGF-β neutralizing antibodies were able to attenuate LV hypertrophy while losartan reduced nonmyocyte proliferation, implying possible therapeutic implications in humans as well [30].

9   NT-proBNP AS A BIOMARKER

NT-proBNP in AOS patients was elevated compared with that in control groups, although none of the patients had extremely high NT-proBNP levels of more than 250 pg/mL [2]. After 1 year, no significant differences in NT-proBNP were observed [7]. In vivo and in vitro studies have shown that treatment with brain natriuretic peptide can attenuate cardiac hypertrophy via the TGF-β1 pathway [31]. One may hypothesize that the elevated NT-proBNP levels in AOS patients are a protective mechanism against the emergence of LV hypertrophy. Because mildly to moderately elevated NT-proBNP levels in other patient groups are reported to predict cardiovascular outcome and AF recurrence, evaluation of the prognostic value of NT-proBNP in AOS patients with respect to clinical outcome may be important [32,33].

10   ARTERIAL STIFFNESS

In the initial Rotterdam cohort of AOS patients, biochemical and arterial stiffness measurements were analyzed in a one-to-one comparison with age-, sex-, and smoking status-matched controls. Results showed that the aortic pulse wave velocity, which is a measure of aortic stiffness, was high-normal in AOS patients [2]. This finding is consistent with previous studies that demonstrated increased aortic stiffness in patients with MFS and BAV [34,35]. Ascending aortic diameter and aortic pulse wave velocity were not correlated, suggesting that arterial stiffness occurs independently of aneurysm formation. The aortic pulse wave velocity did not change after 1 year of follow-up [7]. For future research directions, it would be interesting to evaluate whether aortic stiffness changes over a longer period of time and whether this change is associated with aneurysm formation in both the aorta of medium-size arteries, such as the carotid artery.

11   PREGNANCY

While similar genetic aortopathies, such as MFS and Ehlers-Danlos syndrome, are well known for the vascular complications that can occur in pregnant women, so far this does not seem to be the case in AOS. None of the aortic dissections or other vascular complications in known AOS patients occurred during pregnancy or delivery, although data are relatively scarce [2,3,5].

12   CONCLUSIONS

AOS is an aggressive, inherited, connective tissue disorder characterized by arterial tortuosity, aneurysms, aortic dissections, and osteoarthritis. Aortic root enlargement is the most common cardiovascular finding in our series, but cerebrovascular abnormalities were also present in more than 50% of patients. Furthermore, aneurysms and tortuosity can extend throughout the arterial tree. Aortic dissections most frequently occur in the fourth decade of life, sometimes at relatively small aortic diameters. Progression of known aortic aneurysms can be fast and unpredictable. In addition, mitral valve abnormalities and congenital heart defects are commonly observed.

REFERENCES

[1] van de Laar IM, Oldenburg RA, Pals G, Roos-Hesselink JW, de Graaf BM, Verhagen JM, et al. Mutations in SMAD3 cause a syndromic form of aortic aneurysms and dissections with early-onset osteoarthritis. Nat Genet 2011;43(2):121–6.

[2] van der Linde D, van de Laar IM, Bertoli-Avella AM, Oldenburg RA, Bekkers JA, Mattace-Raso FU, et al. Aggressive cardiovascular phenotype of aneurysms-osteoarthritis syndrome caused by pathogenic SMAD3 variants. J Am Coll Cardiol 2012;60(5):397–403.

[3] van de Laar IM, van der Linde D, Oei EH, Bos PK, Bessems JH, Bierma-Zeinstra SM, et al. Phenotypic spectrum of the SMAD3-related aneurysms-osteoarthritis syndrome. J Med Genet 2012;49(1):47–57.

[4] Loeys BL, Dietz HC, Braverman AC, Callewaert BL, De Backer J, Devereux RB, et al. The revised Ghent nosology for the Marfan syndrome. J Med Genet 2010;47(7):476–85.

[5] Regalado ES, Guo DC, Villamizar C, Avidan N, Gilchrist D, McGillivray B, et al. Exome sequencing identifies SMAD3 mutations as a cause of familial thoracic aortic aneurysm and dissection with intracranial and other arterial aneurysms. Circ Res 2011;109(6):680–6.

[6] Aubart M, Gobert D, Aubart-Cohen F, Detaint D, Hanna N, d'Indya H, et al. Early-onset osteoarthritis, Charcot-Marie-Tooth like neuropathy, autoimmune features, multiple arterial aneurysms and dissections: an unrecognized and life threatening condition. PLoS One 2014;9(5):e96387.

[7] van der Linde D, Bekkers JA, Mattace-Raso FU, van de Laar IM, Moelker A, van den Bosch AE, et al. Progression rate and early surgical experience in the new aggressive aneurysms-osteoarthritis syndrome. Ann Thorac Surg 2013;95(2):563–9.

[8] Salim MA, Alpert BS, Ward JC, Pyeritz RE. Effect of beta-adrenergic blockade on aortic root rate of dilatation in the Marfan syndrome. Am J Cardiol 1994;74(6):629–33.

[9] Kornbluth M, Schnittger I, Eyngorina I, Gasner C, Liang DH. Clinical outcome in the Marfan syndrome with ascending aortic dilatation followed annually by echocardiography. Am J Cardiol 1999;84(6):753–5.

[10] Meijboom LJ, Timmermans J, Zwinderman AH, Engelfriet PM, Mulder BJ. Aortic root growth in men and women with the Marfan's syndrome. Am J Cardiol 2005;96(10):1441–4.

[11] Lazarevic AM, Nakatani S, Okita Y, Takeda Y, Hirooka K, Matsuo H, et al. Determinants of rapid progression of aortic root dilatation and complications in Marfan syndrome. Int J Cardiol 2006;106(2):177–82.

[12] Jondeau G, Detaint D, Tubach F, Arnoult F, Milleron O, Raoux F, et al. Aortic event rate in the Marfan population: a cohort study. Circulation 2012;125(2):226–32.

[13] Ferencik M, Pape LA. Changes in size of ascending aorta and aortic valve function with time in patients with congenitally bicuspid aortic valves. Am J Cardiol 2003;92(1):43–6.

[14] Davies RR, Kaple RK, Mandapati D, Gallo A, Botta DM Jr, Elefteriades JA, et al. Natural history of ascending aortic aneurysms in the setting of an unreplaced bicuspid aortic valve. Ann Thorac Surg 2007;83(4):1338–44.

[15] Thanassoulis G, Yip JW, Filion K, Jamorski M, Webb G, Siu SC, et al. Retrospective study to identify predictors of the presence and rapid progression of aortic dilatation in patients with bicuspid aortic valves. Nat Clin Pract Card 2008;5(12):821–8.

[16] van der Linde D, Yap SC, van Dijk AP, Budts W, Pieper PG, Van der Burgh PH, et al. Effects of rosuvastatin on progression of stenosis in adult patients with congenital aortic stenosis (PROCAS trial). Am J Cardiol 2011;108(2):265–71.

[17] Martens T, Van Herzeele I, De Ryck F, Renard M, De Paepe A, Francois K, et al. Multiple aneurysms in a patient with aneurysms-osteoarthritis syndrome. Ann Thorac Surg 2013;95(1):332–5.

[18] van der Linde D, Verhagen HJ, Moelker A, van de Laar IM, van Herzeele I, De Backer J, et al. Aneurysm-osteoarthritis syndrome with visceral and iliac artery aneurysms. J Vasc Surg 2013;57(1):96–102.

[19] van der Linde D, Witsenburg M, van de Laar I, Moelker A, Roos-Hesselink J. Saccular aneurysm within a persistent ductus arteriosus. Lancet 2012;379(9816):e33.

[20] Chaikof EL, Fillinger MF, Matsumura JS, Rutherford RB, White GH, Blankensteijn JD, et al. Identifying and grading factors that modify the outcome of endovascular aortic aneurysm repair. J Vasc Surg 2002;35(5):1061–6.

[21] Hilhorst-Hofstee Y, Scholte AJ, Rijlaarsdam ME, van Haeringen A, Kroft LJ, Reijnierse M, et al. An unanticipated copy number variant of chromosome 15 disrupting SMAD3 reveals a three-generation family at serious risk for aortic dissection. Clin Genet 2013;83(4):337–44.

[22] van der Linde D, Konings EE, Slager MA, Witsenburg M, Helbing WA, Takkenberg JJ, et al. Birth prevalence of congenital heart disease worldwide: a systematic review and meta-analysis. J Am Coll Cardiol 2011;58(21):2241–7.

[23] Fitzgerald KK, Bhat AM, Conard K, Hyland J, Pizarro C. Novel SMAD3 mutation in a patient with hypoplastic left heart syndrome with significant aortic aneurysm. Case Rep Genet 2014;2014:591516.

[24] Arthur HM, Bamforth SD. TGF-signaling and congenital heart disease: insights from mouse studies. Birth Defects Res A 2011;91(6):423–34.

[25] Armstrong EJ, Bischoff J. Heart valve development: endothelial cell signaling and differentiation. Circ Res 2004;95(5):459–70.

[26] Khan R, Sheppard R. Fibrosis in heart disease: understanding the role of transforming growth factor-beta in cardiomyopathy, valvular disease and arrhythmia. Immunology 2006;118(1):10–24.

[27] Gramley F, Lorenzen J, Koellensperger E, Kettering K, Weiss C, Munzel T. Atrial fibrosis and atrial fibrillation: the role of the TGF-β1 signaling pathway. Int J Cardiol 2010;143(3):405–13.

[28] Alpendurada F, Wong J, Kiotsekoglou A, Banya W, Child A, Prasad SK, et al. Evidence for Marfan cardiomyopathy. Eur J Heart Fail 2010;12(10):1085–91.

[29] Rosenkranz S. TGF-beta1 and angiotensin networking in cardiac remodeling. Cardiovasc Res 2004;63(3):423–32.

[30] Teekakirikul P, Eminaga S, Toka O, Alcalai R, Wang L, Wakimoto H, et al. Cardiac fibrosis in mice with hypertrophic cardiomyopathy is mediated by non-myocyte proliferation and requires TGF-β. J Clin Invest 2010;120(10):3520–9.

[31] He JG, Chen YL, Chen BL, Huang YY, Yao FJ, Chen SL, et al. B-type natriuretic peptide attenuates cardiac hypertrophy via the transforming growth factor-ß1/smad7 pathway in vivo and in vitro. Clin Exp Pharmacol P 2010;37(3):283–9.

[32] den Uijl DW, Delgado V, Tops LF, Ng AC, Boersma E, Trines SA, et al. Natriuretic peptide levels predict recurrence of atrial fibrillation after radiofrequency catheter ablation. Am Heart J 2011;161(1):197–203.

[33] Goei D, van Kuijk JP, Flu WJ, Hoeks SE, Chonchol M, Verhagen HJ, et al. Usefulness of repeated N-terminal pro-B-type natriuretic peptide measurements as incremental predictor for long-term cardiovascular outcome after vascular surgery. Am J Cardiol 2011;107(4):609–14.

[34] Kiotsekoglou A, Moggridge JC, Saha SK, Kapetanakis V, Govindan M, Alpendura F, et al. Assessment of aortic stiffness in Marfan syndrome using two-dimensional and Doppler echocardiography. Echocardiography 2011;28(1):29–37.

[35] Tzemos N, Lyseggen E, Silversides C, Jamorski M, Tong JH, Harvey P, et al. Endothelial function, carotid-femoral stiffness, and plasma matrix metalloproteinase-2 in men with bicuspid aortic valve and dilated aorta. J Am Coll Cardiol 2010;55(7):660–8.

Chapter 3

Systemic Features of Aneurysms-Osteoarthritis Syndrome

I.M.B.H. van de Laar, MD, PhD, M.W. Wessels, MD, PhD

1 JOINT ANOMALIES

Upon discovery of the gene in 2011, our attention was drawn to the osteoarticular phenotype of the *SMAD3* mutation carriers, because in our initial cohort, almost all the patients developed early-onset joint abnormalities, including osteoarthritis and osteochondritis dissecans, meniscal lesions, and intervertebral disc degeneration; additionally, the *Smad3* knockout mice show osteoarthritis and intervertebral disc degeneration [1,2].

Osteoarthritis in the extremities is characterized by the degradation of articular cartilage and subchondral bone of joints. Osteochondritis dissecans is defined as a separation of an articular cartilage and subchondral bone segment from the remaining articular surface. These abnormalities were already present at a young age and were often the patient's presenting symptom. They seem to be discriminating clinical features in AOS.

In our initial report, the mean age at osteoarthritis diagnosis was 42 years, and the youngest patient with osteoarthritis was detected at 12 years of age [3]. The joints that were mostly affected were in the spine, hands and/or wrists, and knees, but osteoarthritis was also reported in all other joints, including feet and/or ankle, hip, and shoulder (Fig. 3.1). The osteoarthritis seen in the hand and wrist only involved the scaphotrapeziotrapezoidal, first carpometacarpal, and, occasionally, metacarpophalangeal joints. In contrast with classical hand osteoarthritis, the distal and proximal interphalangeal joints were not affected. Furthermore, intervertebral disc degeneration mainly involving the cervical and lumbar discs was present in a vast majority of patients upon retrospective evaluation of X-rays and CT scans. In addition, vertebral bodies located in the region of the anterior growth plates showed shape irregularities. These abnormalities were in some documented cases already present at a young age (youngest 12 years). Spondylolysis and/or spondylolisthesis were common. More than half of the patients had nontraumatic osteochondritis dissecans even at a young age. Osteochondritis dissecans occurred mainly in the knee and occasionally in the ankle or hip. Patients with osteochondritis dissecans

Aneurysms-Osteoarthritis Syndrome. http://dx.doi.org/10.1016/B978-0-12-802708-0.00003-X
Copyright © 2017 Elsevier Inc. All rights reserved.

21

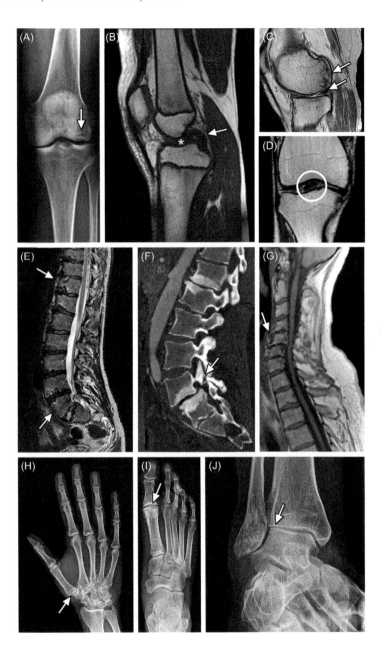

◄ **FIGURE 3.1   Osteoarthritis and osteochondritis.** (A) X-ray of the left knee of a 41-year-old patient shows a large osteochondral lesion without separation in the lateral femoral condyle *(arrow)*. (B) MRI of the right knee of a 12-year-old patient shows congenital absence of the anterior cruciate ligament (ACL). *Asterisk*: no ACL is seen in its expected location. The arrow points to the normal posterior cruciate ligament. (C) MRI of the knee of a 48-year-old woman shows a large osteochondral lesion without separation in the medial femoral condyle *(upper arrow)*. There is also a horizontal tear of the medial meniscus *(lower arrow)*. (D) MRI of the right knee of a 17-year-old man with a loose intraarticular body *(encircled)* due to a large osteochondral lesion of the medial femoral condyle (not shown). (E) MRI of the thoracic and lumbar spine of a 17-year-old man with marked irregularity and impression of the anteroinferior endplates at multiple levels (eg, see arrows). (F) CT scan of a 44-year-old woman with severe multilevel degenerative disc disease and a spondylolisthesis at the L4–L5 level due to a bilateral spondylolysis *(arrow)*. (G) MRI scan of a 50-year-old man with marked abnormalities of the lower cervical spine *(arrow)* with narrowing of the spinal cord. (H) X-ray of the right hand and wrist of a 31-year-old man with moderate osteoarthritis of the first carpometacarpal joint *(arrow)*. (I) X-ray of the right foot of a 31-year-old man with moderate osteoarthritis of the first metatarsophalangeal joint *(arrow)*. (J) X-ray of the ankle of a 40-year-old woman with marked degenerative changes of the talocrural joint with severe lateral joint space narrowing *(arrow)*. *(Reprinted with permission from the article by van de Laar IM et al. Phenotypic spectrum of the SMAD3-related aneurysms-osteoarthritis syndrome. J Med Genet. 2012;49(1):47–57.)*

were operated on before the age of 40 years (youngest 10 years). Osteochondritis dissecans were asymptomatic in some cases. About a quarter of patients had meniscal lesions, one of whom had bilateral meniscectomy at the age of 13 years. In one patient, a congenital absence/agenesis of the anterior cruciate ligament was seen on an MRI of the knees at the age of 12 years. Joint laxity, defined as a Beighton score of ≥5, was seen in a minority (10%) of patients. Joint pain was significantly more frequent in SMAD3 gene mutation carriers as compared to healthy subjects or Marfan syndrome (MFS) patients carrying a FBN1 mutation [4].

The high prevalence of osteoarticular manifestations was confirmed by Mélodie Aubart et al., who studied 50 *SMAD3* mutation carriers and found joint involvement in 100% of their patients [4]. The prevalence of osteochondritis dissecans was significantly lower in their cohort as compared to our study (4% vs 56%) [4].

However, in recent years, several studies have reported individuals with pathogenic *SMAD3* mutations without osteoarticular manifestations, indicating that this finding is not mandatory [5–7]. When we include all published *SMAD3* mutation carriers about whom clinical information on osteoarticular phenotype has been reported, the prevalence of osteoarthritis is still 63%, indicating that osteoarthritis is an important diagnostic clue [4–11].

Joint anomalies, including osteoarthritis, osteochondritis dissecans, and meniscal lesions, are rarely described in patients with LDS, due to *TGFBR1* and *TGFBR2* mutations, or in MFS, but no systematic joint studies in these patients have been reported. Further studies to establish the frequency of osteoarthritis and osteochondritis dissecans in these related syndromes are warranted [12,13].

2   SKELETAL ANOMALIES

A study in 2012 revealed that approximately 40% of AOS patients had long and slender fingers and toes, but overt arachnodactyly was not present [2,3]. Dolichostenomelia was present in 21% of patients. Thorax deformities, including pectus carinatum, pectus excavatum, or asymmetry of the costosternal junction, were present in 36% of patients. Scoliosis was present in 61% of our patients, and three of them were operated on for severe scoliosis. One patient had foraminal stenosis requiring foraminotomy of L5-S1 with spondylodesis of L4-S1. Protrusio acetabulae was present in approximately one-third (35%) of patients and was usually mild. More than 90% of patients had pes planus. Camptodactyly was present in 13% of patients.

Aubart et al. reported scoliosis in 50% of their patients and Scheuermann disease in 22% [4]. In a recent study on patients with idiopathic scoliosis, pathogenic *SMAD3* mutations were identified in 0.6% [14].

3   CRANIOFACIAL ABNORMALITIES

The facial features in AOS patients included high forehead, hypertelorism, long face, flat supraorbital ridges, and malar hypoplasia, but they were generally mild [2,3]. Uvular anomalies (raphe, broad, or bifid) were present in more than half of the patients (52%) and mainly included broad uvulas with or without a raphe. Uvular abnormalities may be an easy diagnostic clue, as they only occurred in patients with LDS but not in patients with other syndromic or non-syndromic forms of thoracic aortic aneurysms and dissections. High-arched palates were common, and one patient was operated on for a cleft palate. Dental malocclusion and retrognathia were occasionally seen. No craniosynostosis has been reported so far. There was a marked inter- and intrafamilial variability in facial features.

In recent papers, craniofacial findings are occasionally reported [8,11]. This infrequency might indicate that craniofacial anomalies are less common or milder than initially reported.

4   NEUROLOGICAL FEATURES

Apart from joint, skeletal, and craniofacial abnormalities, Aubart et al. described *SMAD3* mutations also associated with neurological features in 15 of 22 patients (68%), such as muscle cramps, paresthesia, hypoesthesia, or gait disturbance [4]. Of these 15 patients, 9 displayed a peripheral neuropathy, and in half of them, the patients' electromyography showed an axonal motor and sensory neuropathy consistent with a Charcot-Marie-Tooth type 2–like neuropathy. In our initial cohort of 45 patients, no neurological symptoms were exhibited; however, no systematic questionnaires on this topic and no detailed neurological examinations were performed [3].

5   IMMUNOLOGICAL FEATURES

Aubade et al. described some immunological features, including allergic manifestations, especially asthma and allergic conjunctivitis; these were present in 11 of 22 (50%) patients. In 36% of patients, autoimmune features, such as

Sjögren's disease, rheumatoid arthritis, and Hashimoto's disease, were described. A recent case report also suggested an association between rheumatoid arthritis and *SMAD3* [11].

6 ADDITIONAL FEATURES

Some features that are common in connective tissue disorders are also frequent in AOS [2,3]. Umbilical and/or inguinal hernia were present in 43% of patients. Pelvic floor prolapse, involving the uterus, bladder, and bowel, occurred in 41% of adult women at a mean age of 50 years (range 43–64). Varices or thread veins were reported or observed in 58%, usually already present at a young age (youngest patient 17 years) and were therapy-resistant. Velvety skin (62%) and striae (53%) were present in the majority of patients. Other recurrent findings included easy bruising and atrophic scars. Recurrent and severe headaches or migraines were present in half of the patients and did not co-occur with the cerebrovascular abnormalities. Almost 40% of patients complained of chronic or intermittent increased fatigue. Ophthalmologic examination in several cases revealed no ectopia lentis, but cataracts were reported sporadically. The prevalence of acrocyanosis was 52%, significantly greater in *SMAD3* gene mutation carriers as compared to healthy subjects or MFS patients [1].

Some additional features occurred sporadically but were not systematically evaluated in all patients, including diverticulosis, dural ectasia, severe lung emphysema, and xanthelasmata. No moderate or severe developmental delay was reported in any patient.

In our experience, all patients with a *SMAD3* mutation exhibited symptoms or signs of AOS, even in childhood. The penetrance of the mutations was nearly 100%, although the expression varied from very mild (isolated bifid uvula) to severe (aortic aneurysm at 12 months of age) [3,5]. Age-dependent progression of the phenotype was evident, as aneurysms and osteoarthritis were encountered mainly during adulthood. Intrafamilial variability was significant, as illustrated by the clinical findings in the large family of 33 AOS patients [3]. So far, no clear genotype-phenotype correlation has been established.

7 CLASSIFICATION OF ANEURYSMS-OSTEOARTHRITIS SYNDROME

Because *SMAD3* mutation carriers show cardiovascular, craniofacial, cutaneous, and skeletal features similar to those of LDS patients, it was suggested to name the syndrome LDS type III. Patients with LDS show more widespread and/or aggressive vascular disease when compared patients with MFS or thoracic aortic aneurysm and dissection, irrespective of the severity of systemic features; and therefore McGarrick et al. proposed a revised nosology stating that there are no specific clinical criteria for the diagnosis of LDS. The presence of mutations in the *TGFBR1*, *TGFBR2*, *SMAD3*, or *TGFB2* gene in combination with documented aneurysm or dissection or a family history of documented LDS should be sufficient to establish the diagnosis of LDS in a patient [15].

REFERENCES

[1] Li CG, Liang QQ, Zhou Q, Menga E, Cui XJ, Shu B, et al. A continuous observation of the degenerative process in the intervertebral disc of SMAD3 gene knock-out mice. Spine (Phila Pa 1976) 2009;34(13):1363–9.

[2] van de Laar IM, Oldenburg RA, Pals G, Roos-Hesselink JW, de Graaf BM, Verhagen JM, et al. Mutations in SMAD3 cause a syndromic form of aortic aneurysms and dissections with early-onset osteoarthritis. Nat Genet 2011;43(2):121–6.

[3] van de Laar IM, van der Linde D, Oei EH, Bos PK, Bessems JH, Bierma-Zeinstra SM, et al. Phenotypic spectrum of the SMAD3-related aneurysms-osteoarthritis syndrome. J Med Genet 2012;49(1):47–57.

[4] Aubart M, Gobert D, Aubart-Cohen F, Detaint D, Hanna N, d'Indya H, et al. Early-onset osteoarthritis, Charcot-Marie-Tooth like neuropathy, autoimmune features, multiple arterial aneurysms and dissections: an unrecognized and life threatening condition. PLoS One 2014;9(5):e96387.

[5] Wischmeijer A, Van Laer L, Tortora G, Bolar NA, Van Camp G, Fransen E, et al. Thoracic aortic aneurysm in infancy in aneurysms-osteoarthritis syndrome due to a novel SMAD3 mutation: further delineation of the phenotype. Am J Med Genet A 2013;161A(5):1028–35.

[6] Regalado ES, Guo DC, Villamizar C, Avidan N, Gilchrist D, McGillivray B, et al. Exome sequencing identifies SMAD3 mutations as a cause of familial thoracic aortic aneurysm and dissection with intracranial and other arterial aneurysms. Circ Res 2011;109(6):680–6.

[7] Martens T, van Herzeele I, de Ryck F, Renard M, de Paepe A, Francois K, et al. Multiple aneurysms in a patient with aneurysms-osteoarthritis syndrome. Ann Thorac Surg 2013;95(1):332–5.

[8] Hilhorst-Hofstee Y, Scholte AJHA, Rijlaarsdam MEB, van Haeringen A, Kroft LJ, Reijnierse M, et al. An unanticipated copy number variant of chromosome 15 disrupting SMAD3 reveals a three-generation family at serious risk for aortic dissection. Clin Genet 2013;83(4):337–44.

[9] Fitzgerald KK, Bhat AM, Conard K, Hyland J, Pizarro C. Novel SMAD3 mutation in a patient with hypoplastic left heart syndrome with significant aortic aneurysm. Case Rep Genet 2014;2014:591516.

[10] Panesi P, Foffa I, Sabina S, Ait Ali L, Andreassi MG. Novel TGFBR2 and known missense SMAD3 mutations: two case reports of thoracic aortic aneurysms. Ann Thorac Surg 2015;99(1):303–5.

[11] Berthet E, Hanna N, Giraud C, Soubrier M. A case of rheumatoid arthritis associated with SMAD3 gene mutation: a new clinical entity? J Rheumatol 2015;42(3):556.

[12] Law C, Bunyan D, Castle B, Day L, Simpson I, Westwood G, et al. Clinical features in a family with an R460H mutation in transforming growth factor beta receptor 2 gene. J Med Genet 2006;43(12):908–16.

[13] Grahame R, Pyeritz RE. The Marfan syndrome: joint and skin manifestations are prevalent and correlated. Brit J Rheumatol 1995;34(2):126–31.

[14] Haller G, Alvarado DM, Willing MC, Braverman AC, Bridwell KH, Kelly M, et al. Genetic risk for aortic aneurysm in adolescent idiopathic scoliosis. J Bone Joint Surg Am 2015;97(17):1411–7.

[15] MacCarrick G, Black JH JIII, Bowdin S, El-Hamamsy I, Frischmeyer-Guerrerio PA, Guerrerio AL, et al. Loeys-Dietz syndrome: a primer for diagnosis and management. Genet Med 2014;16(8):576–87.

Chapter 4

Differential Diagnosis in Heritable Thoracic Aortic Diseases

D. van der Linde, MD, MSc, PhD, J.W. Roos-Hesselink, MD, PhD

1 INTRODUCTION

Heritable thoracic aortic aneurysms and dissections often occur at a younger age when compared to atherosclerotic disease. Multiple family members can be affected via various inheritance forms, with variable ages of onset and penetrance rates. Heritable thoracic aortic diseases are subdivided into nonsyndromic forms, which can be associated with, for example, a bicuspid aortic valve, and syndromic forms, with prominent features of connective tissue disorders, such as Marfan and Loeys-Dietz syndromes. The syndromic forms are often caused by genes encoding for proteins involved in the transforming growth factor-beta (TGF-β) signaling pathway.

When a patient presents with an aortic aneurysm or dissection at a relatively young age, it is critical to determine which form of heritable thoracic aortic disease is the cause, because diagnostic and treatment guidelines differ per diagnosis. However, because many clinical features overlap between the various heritable thoracic aortic diseases, detailed knowledge about the clinical features and genetics is essential for the correct diagnosis. In this chapter, the following five disease entities, which are well-known causes of heritable thoracic aortic disease are discussed:

- Marfan syndrome
- Loeys-Dietz syndrome
- Ehlers-Danlos syndrome
- Bicuspid aortic valve
- Turner syndrome

Each subchapter aims to provide an overview of its respective disease by discussing the diagnostic criteria, genetic aspects, pathophysiology, clinical manifestations, and management and treatment. The overlap and differences between the specific disease entities are also discussed.

Aneurysms-Osteoarthritis Syndrome. http://dx.doi.org/10.1016/B978-0-12-802708-0.00004-1
Copyright © 2017 Elsevier Inc. All rights reserved.

Chapter 4a

Marfan Syndrome

J. de Backer, MD, PhD

1 DEFINITIONS AND DIAGNOSIS

Marfan syndrome (MFS) is a heritable connective tissue disorder with multi-systemic manifestations, caused by mutations in the fibrillin-1 gene (*FBN1*) [1]. Historical estimates of the prevalence report figures of 1 in 5,000 affected individuals in the general population [2]. More recent epidemiological studies document a wider range of the prevalence of MFS as being between 1.459 and 17.2 per 100,000 persons, with no gender, ethnic, or racial predilection [3]. Of note, conventional prevalence estimates in the literature so far reflect the "phenotype prevalence," based on clinical presentations. A recent study assessing the prevalence of reported *FBN1* variants in the ESP database identified a "genotype prevalence" of 1 in 65. Because the penetrance of *FBN1* mutations is very high, this proportion seems to be an overestimation, which may indicate that at least a proportion of the currently reported variants needs careful reassessment [4].

In his 1956 monograph *Heritable Disorders of Connective Tissue*, Victor McKusick first established a classification of connective tissue disorders, including MFS [5]. In 1986, an expert panel established the first international nosology of heritable disorders of connective tissue with the aim of improving communication between the medical community, patients, and researchers [6]. This so-called Berlin nosology was revised after the identification of the causal gene of MFS in 1996, [7] at which point it was renamed the Ghent nosology. At the 7th International Marfan meeting in Ghent in 2006, a need for revision of the nosology was expressed. A subsequent expert panel meeting led to a revised Ghent nosology, emphasizing the need to correctly identify patients at risk for aortic aneurysm or dissection with criteria that are user-friendly; allow early diagnosis; consider the availability and costs of diagnostic tests; better define entities, such as familial ectopia lentis, MASS phenotype (Myopia, Mitral valve prolapse [MVP], mild Aortic dilatation, Skeletal features, Striae), and MVPS (Mitral Valve Prolapse Syndrome); and delineate triggers for alternative diagnoses, such as Loeys-Dietz syndrome [8]. Aortic root dilatation and ectopia lentis are the cardinal diagnostic features of MFS in this revised nosology. A systemic

Aneurysms-Osteoarthritis Syndrome. http://dx.doi.org/10.1016/B978-0-12-802708-0.00005-3
Copyright © 2017 Elsevier Inc. All rights reserved.

29

TABLE 4A.1 Summary of the Revised Ghent Nosology (Upper Panel) and Systemic Score (Lower Panel)

In the absence of a family history, the diagnosis of MFS is confirmed when either

Ao (Z ≥ 2) + EL
Ao (Z ≥ 2) + *FBN1*
Ao (Z ≥ 2) + Syst (≥7pts)
EL + *FBN1* with known Ao

In the presence of a family history, the diagnosis of MFS is confirmed when either

EL + FH of MFS
Syst (≥7 pts) + FH of MFS
Ao (Z ≥ 2 in adults, Z ≥ 3 in children) + FH of MFS

Systemic score

Pectus carinatum deformity—*Pectus excavatum or chest asymmetry*	**2 − 1**
Wrist AND thumb sign—*wrist OR thumb sign*	**3 − 1**
Hindfoot deformity—*Plain pes planus*	**2 − 1**
Protrusio acetabuli	**2**
Pneumothorax	**2**
Dural ectasia	**2**
Reduced US/LS AND increased arm/height AND no severe scoliosis	**1**
Scoliosis or thoracolumbar kyphosis	**1**
Reduced elbow extension	**1**
Facial features (3/5)	**1**
Skin striae	**1**
Myopia > 3 diopters	**1**
Mitral valve prolapse (all types)	**1**
Total	**20 points**

Ao (Z ≥ 2), Aortic root Z-score at the level of the sinuses of Valsalva exceeding 2 standard deviations above the mean; EL, ectopia lentis; Syst, systemic score; FBN1, documented mutation in the *FBN1* gene; FH, family history.
Source: Adapted from Ref. [8].

score based on the identification of manifestations in different organ systems is added and is considered positive when ≥7/20 (Table 4a.1).

Genetic testing of the *FBN1* gene, although not mandatory, has greater weight in the diagnostic assessment compared to the previous nosology. The criteria according to the revised Ghent nosology are summarized in Table 4a.1. Criteria are less stringent in the presence of a family history of MFS.

Fig. 4a.1 is an illustration of the major skeletal features of MFS.

A critical issue and possible weakness in the revised nosology is that it requires correct measurement of the aortic root dimensions, where z-scores have

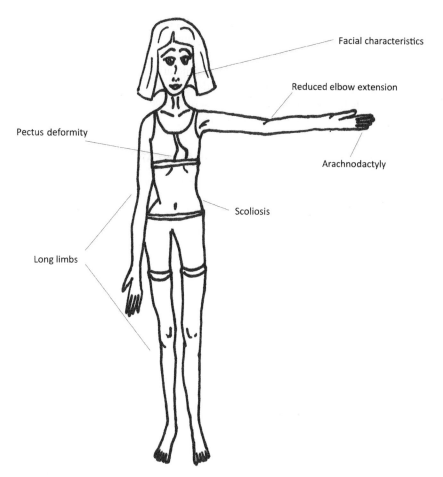

Facial characteristics

Reduced elbow extension

Pectus deformity

Arachnodactyly

Scoliosis

Long limbs

FIGURE 4A.1 Typical clinical manifestations in Marfan syndrome.

to be calculated to express the degree of deviation from the norm. A z-score is defined as the number of standard deviations an observed aortic diameter is above or below the regression line for the body surface area of an individual. Recent studies have indicated that gender should also be taken into account for z-score calculation [9,10]. A z-score > +2 is commonly used to define dilatation and represents a measurement greater than two standard deviations above the mean for age, sex, and body surface area. When the z-score exceeds +3, the term "aneurysm" is commonly applied. Some important issues regarding z-score calculation merit brief mention: (1) the method used to calculate the z-score should be exactly the same as the one used to generate the reference values used in the calculation—such aspects as imaging technique, location of the measurement, and timing are crucial; (2) the reference population should match and cover the age range you are dealing with; and (3) without any change in the aortic diameter, the interpretation of whether it is dilated in the same patient can vary as the patient grows or gains or loses weight [11]. For echocardiographic measurements made from inner wall to inner wall during systole in individuals ≤25 years, a z-score calculator can be found at http://www.marfan.org. For echocardiographic measurements obtained using the leading edge to leading edge technique in diastole in all age groups, reference graphs and z-score equations have recently been published [9,10,12].

2   CLINICAL GENETIC ASPECTS OF MARFAN SYNDROME

MFS is inherited as an autosomal dominant trait. Since the identification of the first *FBN1* mutation in 1991, [13] no fewer than 1847 different mutations in 3044 DNA samples have been reported throughout the gene, according to the currently most extensive *FBN1* mutation database (http://www.umd.be/FBN1/; last updated 8/28/2014). Approximately 12% of all reported *FBN1* mutations are recurrent [14]. Because not all mutations that have been identified worldwide have been reported or entered into the database, it is very likely that the actual numbers are much higher.

Penetrance of *FBN1* mutations in MFS is extremely high, and no example of nonpenetrance has been documented in hundreds of pedigrees [15]. About 25% of MFS cases are caused by new or spontaneous mutations arising during zygote formation. Gonadal mosaicism has been reported sporadically in patients with MFS [16,17].

Evidence has shown that paternal age at the time of conception of isolated cases of MFS is advanced (36 vs 29 years), consistent with the knowledge that a new mutation in a spermatagonium is a frequent cause of sporadic MFS [18].

The extensive variation in characteristics of the mutations and the known clinical variability in MFS have prompted many researchers to explore genotype-phenotype correlations. Correlations of various phenotypic features with regard to location within the gene and type of mutation have been explored and reported, [19–27] but application in clinical practice has not been evidenced for most of them. Cysteine missense mutations more often lead to ectopia lentis, and the substitution of a cysteine residue leads to a more severe cardiovascular and skeletal phenotype than the introduction of a cysteine residue [24]. Nonsense and splicing variants were recently found to occur more frequently in MFS patients with aortic events [27]. The established marked phenotypic variability among individuals carrying the same mutation tends to blur distinctions among different mutations, [28] and it is likely that epigenetic and environmental factors also play an important role.

The most robust genotype-phenotype correlation is for point mutations or small in-frame deletions in the epidermal growth factor–like motifs in the middle region of the gene (exons 24–32), which tend to result in a more severe form of the disorder and explain virtually all cases of neonatal MFS [23]. One case of neonatal MFS has been reported with a homozygous mutation (two mutations within six base pairs from each other in exon 26) [29].

Other factors to explain the high clinical variation in MFS that have been explored and are the subject of ongoing research include the level of expression of the normal *FBN1* allele [30,31] and hyperhomocysteinemia [32].

The clinical variation in *FBN1* carriers extends even beyond the strict diagnosis of MFS—indeed, mutations have been identified in patients presenting milder phenotypes of phenotypes restricted to one organ system [33]. These milder Marfan-related phenotypes include MASS phenotype (Myopia, MVP, mild Aortic dilatation, Skeletal features, Skin striae) (MIM# 604308; OR-PHA#99715), [34] familial ectopia lentis (MIM# 129600; ORPHA#1885), familial kyphoscoliosis, [35] and predominant thoracic aortic aneurysms (MIM# 132900; ORPHA#91387) [36,37]. Importantly, patients presenting these "milder" phenotypes need careful and extended clinical follow-up, because some features, such as aortic dilatation, may still become apparent after many years. Also notable in this context is that some patients who were previously classified as one of these milder phenotypes may be reclassified when the new diagnostic criteria for MFS are applied [8]. Indeed, patients not meeting the clinical criteria themselves but harboring a *FBN1* mutation known to be associated with MFS in other families should be considered as MFS, as has been reported in the case of isolated MFS by Aman Chandra et al. [38]. The importance of the impact of the causality of an identified *FBN1* variant in the diagnosis stresses the urgent need for more extended and well-curated *FBN1* databases.

Interestingly, not all *FBN1* mutations lead to MFS or even to aortic disease. The clinical features of these allelic disorders are quite distinct and, in some cases, even the opposite of classical MFS features. Acromicric dysplasia and geleophysic dysplasia are characterized by short stature, short hands, and stiff joints. These skeletal dysplasias are caused by mutations in the TB5 domain of the *FBN1* gene [39]. A deletion in the TB5 domain had been previously reported in Weill-Marchesani syndrome, another entity characterized by short stature, short hands, and stiff joints that differs from acromicric dysplasia and geleophysic dysplasia by the presence of microspherophakia [40]. A distinct disorder, stiff skin syndrome is caused by mutations in the TB4 domain [41]. Patients with stiff skin syndrome do not present with short stature or short hands but have a stony-hard and thickened skin that results in joint contractures. Some patients have a "hybrid" phenotype with ectopia lentis, and thus aortic and ophthalmological follow-up is recommended.

Finally, mutations in exon 29 of the *FBN1* gene have been reported in two patients with Shprintzen-Goldberg syndrome (SGS), the only other *FBN1*-related disorder known to associate with aortic aneurysms, albeit restricted to a minority of patients. SGS is further characterized by a combination of distinctive facial features (craniosynostosis and proptosis), skeletal manifestations, and intellectual disability [42]. The major gene mutated in SGS, however, is SKI (see later). The exact relationship between these specific phenotypes and the underlying *FBN1* mutation needs further investigation but clearly illustrates the complex properties of fibrillin-1 in the extracellular matrix.

3 CLINICAL MANIFESTATIONS

3.1 Cardiovascular Manifestations

3.1.1 Aorta and Pulmonary Artery

Most of the morbidity and mortality associated with MFS is related to the cardiovascular manifestations. The most common lesion is dilatation at the level of the sinuses of Valsalva, leading to aneurysm formation, which can ultimately be complicated by aortic dissection or rupture when left untreated or not identified in an early stage. It is estimated that aortic root dilatation is present in >80% of adult MFS patients [1,43–48]. Despite the availability of prophylactic replacement of the aortic root, even current surgical series document that up to one-third of MFS patients present with aortic dissection, mainly Stanford type A dissection. Such high numbers of acute aortic emergencies indicate the persisting deficiency of timely diagnosis and adequate risk estimation [1,49].

The ascending aorta may be the weak spot in MFS because of structural and local hemodynamic factors being typical for that part of the aorta. As McKusick described so accurately in 1956, disease manifestations in MFS, including vascular disease, should be considered as an abiotrophy, where the involved tissues wear out prematurely under the usual stresses and strains [50]. Structural abnormalities in the ascending aorta arise from the complex and not fully unraveled interplay between embryologic factors, abnormalities in the elastic fiber formation and homeostasis, and altered signaling pathways, including the TGF-β pathway. The embryologic development of the aorta is complex, with different parts being developed at different times and stemming from a different origin, as nicely summarized in a recent review by Hisham Sherif [51]. Heterogeneity of embryological origins is a hallmark of aortic vascular smooth muscle cells (VSMCs) [6,52]. VSMCs of the aortic root arise from the secondary heart field, a lateral plate mesoderm derivative; the ascending aorta and arch are derived from neural crest cells; the supradiaphragmatic part of the descending aorta is derived from somitic mesoderm; and the abdominal aorta is derived from splanchnic mesoderm. Interestingly the sites of developmental field boundaries seem to correspond anatomically to sites of predisposition for aneurysms and dissections (the aortic and pulmonary roots, the ductal region of the aorta, and the suprarenal abdominal aorta) [53,54]. So far, there seems to be no evidence for a generalized effect on the development of specific aortic segments by one single gene, receptor, or pathway. Notable in the context of this review, the effect of TGF-β on the expression of smooth muscle cell genes is different according to the location in the aorta. Different isoforms of TGF-β exist and seem to exhibit a different effect on smooth muscle cell transcriptional responses according to the lineage of the cells (ectodermal vs mesodermal). Similar observations have been made for other important genes in this context, including smooth muscle α-actin and myosin heavy chain (*MYH11*) [51]. It has been demonstrated that the elastic fiber content is higher in the ascending aorta than in any other region of the arterial tree [55]. Such diseases as MFS that affect elastic fiber integrity therefore manifest more easily in this region. Furthermore, it is primarily the ascending portion of the aorta that is subjected to the repetitive stress of left ventricular (LV) ejection, eventually resulting in progressive dilatation [9,10,12,56–60].

Because pressures in the aorta are significantly higher than in the pulmonary artery, dilatation is more pronounced at the aortic root. Pulmonary artery dilatation occurs commonly in patients with MFS, but it rarely leads to dissection or rupture [58,61–64]. Cut-off values for the assessment of pulmonary artery dilatation are available for CT/MRI and for echocardiography [24,27,63,65].

Up to now, complications in the descending aorta have been limited to a minority of MFS patients. However, the incidence of descending thoracic aortic dissection may increase with the growing age of MFS patients in whom only the ascending aorta has been replaced [66,67]. MFS patients having thoraco-abdominal aortic aneurysm/dissection as a presenting manifestation are discussed in a few case reports, [68,69] and it is notable that dissection in the descending part of the aorta is independent of the diameter of the ascending aorta [70]. Other data on the descending aorta in MFS patients are found in surgical reports describing the occurrence of primary or secondary complications in the descending aorta, necessitating surgical intervention. Among MFS patients, 8–15% require initial surgery in the descending aorta [42,49,66]. Patients with initial type B aortic dissection are at a significant higher risk for reintervention (86% for previous type B dissection versus 42% for previous type A dissection). The majority of reinterventions is required in patients with previous dissection (48 vs 11% reintervention in the patients presenting with aortic aneurysm) [49]. A large contemporary series of 96 thoraco-abdominal aortic aneurysm repairs in patients with MFS shows an excellent survival rate of 97% [71]. Recommendations are slightly different from the ones applied for the aortic root (see later): repair at the level of the thoraco-abdominal in MFS is recommended when the aneurysm diameter exceeds 5.5 cm [72].

Altered anatomical features in the aorta of MFS patients, such as dilatation and/or dissection, are accompanied by functional impairment of the vessel, as reflected in increased aortic stiffness in the patients. Abnormal elastic properties—pulse wave velocity and local aortic distensibility—are not confined to the ascending aorta but also detected in the normal-size, more distal parts of the vessel [53,73]; this applies to patients who have previously undergone aortic root surgery as well as to MFS patients who have not been operated upon [74]. Interestingly, aortic stiffness seems to be an early marker of aortic disease, as MFS patients already demonstrate significant decreased aortic distensibility at proximal as well as distal levels of the aorta prior to dilatation of the thoracic aorta [53]. Local distensibility of the descending thoracic aorta appears to be an independent predictor of progressive descending aortic dilatation [75]. Interestingly, increased augmentation index as an indirect marker of aortic stiffness has been found to predict progression of aortic disease in MFS independently of aortic root size [76]. Major determinants of the augmentation index are pulse wave velocity and the length of the arterial tree. Alterations in wave reflections seem to be more pronounced in younger MFS patients [77]. Although these data on functional alterations of the aorta are definitely worthy of further studies, they have not yet reached the level of clinical application. Larger-scale studies are required to assess their validity in models for more robust risk stratification in MFS.

3.1.2 Mitral Valve Prolapse

MVP is another common cardiovascular complication of MFS, occurring in up to 80% of patients [64,78,79]. It is unclear whether the prevalence of MVP is increasing with age. In a large pediatric cohort, MVP occurred more commonly in females, but this female preponderance was not confirmed in adult series [80,81]. A recent study showed that MFS patients have an increased risk of mitral valve–related clinical events at a younger age (endocarditis, surgery, and heart failure) when compared to patients with an idiopathic MVP (28 vs 13%, respectively) [12]. Among MFS patients, 5–12% will require mitral valve surgery as a primary cardiovascular intervention [82].

3.1.3 Cardiomyopathy and Arrhythmias

Although it is not commonly acknowledged in MFS, dilated cardiomyopathy, beyond that explained by aortic or mitral valve regurgitation, seems to occur with a higher prevalence, suggesting a role for the extracellular matrix protein fibrillin-1 in the myocardium. Significant LV dilatation and dysfunction leading to heart failure and necessitating heart transplantation have been described in a few cases and seem to be a very rare complication [83]. Subclinical myocardial dysfunction, on the other hand, has been reported in larger subsets of MFS patients of various ages by several independent research groups, [43–48] and mildly increased LV dimensions have been demonstrated in a subset of patients with MFS [84]. The presence of mild intrinsic cardiomyopathy has recently been confirmed in a mouse model of MFS ($fbn1^{C1039G/+}$) [85]. A study assessing genotype-phenotype correlations indicated that LV dilatation was more frequently observed in patients with a nonmissense *FBN1* mutation [86]. Recently, studies in another mouse model for MFS ($fbn1^{mgR/mgR}$) provided interesting new insights into the pathophysiology of MFS cardiomyopathy, indicating altered muscle mechanosignaling as a trigger [87].

A feature that is closely related to ventricular dysfunction in MFS is an increased risk for adverse arrhythmogenic events, as evidenced by several groups [56,58,60,63]. Boris Hoffmann et al. found an association with increased NT-proBNP levels [58,65].

An illustration of the main cardiovascular features in MFS is provided in Fig. 4a.2.

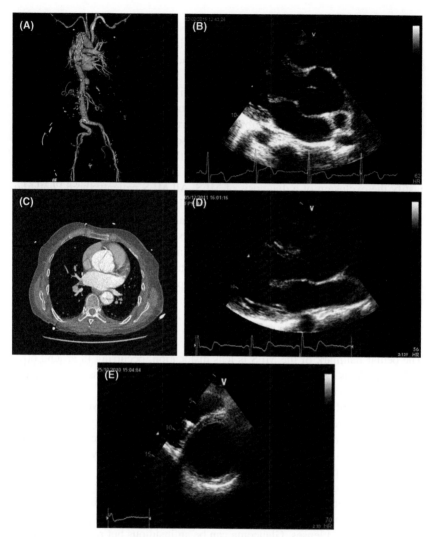

FIGURE 4A.2   Cardiovascular manifestations of Marfan syndrome. (A) CTA image of a descending thoracic aortic aneurysm in a MFS lady with a history of replacement of the ascending (Bentall), descending thoracic and abdominal aorta—the aneurysm is located on a small segment of remaining native aorta—also note the totuous iliac arteries and the stent in the subclavian artery. (B) TTE image of a sinus of Valsalva aneurysm. (C) CTA image of a type A dissection in a MFS lady with a diameter of the aortic root at the sinus of Valsalva of 45 mm on echocardiography 3 months prior to the dissection—the dissection was initiated at the right coronary artery branching point and extended into the iliac arteries. (D) Abnormal thickening of the anterior mitral valve leaflet in a MFS patient with Mitral valve prolapse. (E) Dilated cardiomyopathy.

3.2 Skeletal Manifestations

Skeletal manifestations in MFS are highly variable with regard to age of onset and severity. They often begin in childhood and can lead to incapacitating scoliosis, pectus deformities, and joint luxations. Skeletal manifestations may result from joint laxity (flat feet, joint hypermobility), long bone overgrowth (tall stature, increased arm span, arachnodactyly), or a combination of both (pectus deformities). Skeletal manifestations are important diagnostic triggers, as many patients have skeletal manifestations as a presenting symptom [88]. Of the 13 features listed in the systemic score, 8 used for the diagnosis of MFS are skeletal features [8]. On the other hand, the absence of skeletal manifestations does not justify the exclusion of a diagnosis of MFS. All MFS patients should undergo regular orthopedic exams, especially during puberty. In adulthood, degenerative arthritis may become an issue due to long-standing joint hypermobility [89]. Severe scoliosis, often in combination with pectus excavatum, may result in restrictive pulmonary disease and even cor pulmonale.

Another underrecognized skeletal manifestation of MFS is decreased bone density, [90,91] which may explain at least in part the chronic pain observed in many MFS patients.

3.3 Ocular Manifestations

The ocular hallmark of the MFS is lens dislocation, occurring in around 60% of MFS patients [92]. Extraction is usually not necessary, unless the dislocated lens interferes with normal vision.

The globe of the eye is often elongated in patients with MFS, leading to myopia and increased stretching of the retina, with a propensity for retinal detachment. A degree of myopia exceeding 3 diopters is part of the systemic score in the revised Ghent nosology. In a recent comprehensive study of ocular findings in 87 MFS patients, myopia >3D occurred in 38% [93].

A flattened cornea was used in the original Ghent diagnostic criteria but was abandoned in the revised nosology in view of the difficulties in assessing this feature. Development of cataracts is accelerated in patients with MFS, especially in displaced lenses. Glaucoma can be an insidious but rare complication of lens dislocation [93].

Other common features include strabismus (especially exotropia), which may lead to amblyopia when left uncorrected.

3.4 Pulmonary Involvement in Marfan Syndrome

Various pulmonary problems are associated with MFS, including restrictive pulmonary disease related to thoracic deformities and apical blebs incurring an increased risk for pneumothorax [94–96]. Impairment of distal alveolar septation has been observed in a mouse model for MFS showing increased TGF-β signaling. In this study, the pulmonary phenotype was reversible with TGF-β antagonists [97].

Obstructive and central sleep apnea occurs with a higher frequency in patients with MFS and is related to cardiovascular parameters, including LV ejection fraction, atrial fibrillation, and mitral valve surgery [98]. Obstructive sleep apnea syndrome appears to be related to aortic events in MFS and deserves further attention in the search for risk stratification for aortic disease in MFS [99].

4 ETIOLOGY AND PATHOPHYSIOLOGY

The pleiotropic clinical manifestations observed in MFS strongly suggested that the causal factor for the disease had to be found in a widespread tissue component, such as the connective tissue. Applying monoclonal antibody studies in mice, David Hollister, Lynn Sakai, and their colleagues discovered a putative candidate localized to numerous tissues, including the aortic media and ciliary zonule, with the fibrils visualized termed fibrillin [100]. Fibrillin fibrils are a part of the 10-nm elastin-associated microfibrils. Immunohistochemical studies were the first to suggest that fibrillin was implicated in MFS [101]. The ultimate proof had to await the discovery of the gene encoding fibrillin-1 (*FBN1*) and subsequent genetic linkage and mutation studies [13,102]. To date, more than 1000 different mutations have been identified throughout the gene, with most being unique to an individual or a family.

The histological hallmark in the aorta of patients with MFS is the marked and premature degeneration of the medial layer, with degeneration of the elastic fibers, irregular hypertrophy and apoptosis of VSMCs, and increased basophilic ground substance within cell-depleted areas. However, aortopathy in patients with MFS is not limited to the aortic media. Indeed, evidence suggests that endothelial cells may also play a role in the pathogenesis. Impaired flow-mediated endothelial response was evidenced a while ago in patients with MFS and suggested a role for fibrillin in endothelial cell mechanotransduction [103]. Another mechanism by which the endothelium may be involved in MFS is through a loss of basal NO production, as shown in an MFS mouse model [104]. Clearly, the role of the endothelium in aneurysm formation in patients with MFS is not elucidated yet, and further studies are definitely needed.

The current knowledge of the role of fibrillin-1 in the pathogenesis of aortic aneurysm formation is at least trifold: (1) structural role in elastic fiber composition, (2) regulator of TGF-β signaling, and (3) role in mechanotransduction.

Classically, aneurysm formation in MFS is considered to result from an inherent *structural weakness* of connective tissues.

Studies in different mouse models for MFS have extended this view by demonstrating that fibrillin-1 also plays an important functional role in *regulating transforming growth factor-beta* (TGF-β) bioavailability. TGF-β is a multifunctional peptide that controls proliferation, differentiation, and other functions in many cell types. Fibrillin-1 is homologous to the family of latent TGF-β binding proteins (LTBPs), which serve to hold TGF-β in an inactive complex in various tissues, including the extracellular matrix [105]. Indeed, fibrillin-1 was shown to bind TGF-β and LTBPs [106–108]. Hence, it was hypothesized that mutations in fibrillin-1 could lead to perturbed sequestration of the inactive TGF-β complex [97]. Indeed, increased TGF-β signaling has been demonstrated in several tissues in MFS patients and murine models for MFS.

Surprisingly, more recent studies demonstrated that a mouse in which the LTBP binding site was deleted ($Fbn1^{H1\Delta}$) did not present features of MFS [109]. This observation refuted the importance of TGF-β sequestration by fibrillin-1, and an alternative hypothesis was proposed whereby mutant microfibrils influence TGF-β activation in a different way. Increased TGF-β signaling is now considered to be the result of a final common pathway in the disease process. The role of the TGF-β signaling pathway may also vary during the dynamic transition from aortic aneurysm predisposition to end-stage disease, such as dissection [110].

Moreover, fibrillins do not only contribute to cell signaling in the vessel wall through regulation of growth factor bioavailability; they are also important in *mechanotransduction* from the endothelium and extracellular matrix to the VSMC. The process of mechanotransduction is critical to maintain homeostasis within the aortic wall by regulating aortic remodeling in response to hemodynamic stress. Mutations in fibrillin-1 may perturb this mechanism [111]. Hence, a recent hypothesis states that the mechanical state of the matrix is sensed by cells in the vessel wall, which consequently send a signal through integrins and the cytoskeleton, resulting in inappropriate remodeling and aneurysm formation via a common pathway of inappropriate TGF-β signaling [111,112]. A schematic overview of the pathogenesis is provided in Fig. 4a.3.

5   MANAGEMENT AND TREATMENT OF MARFAN SYNDROME

The pleiotropic nature of MFS implies a multidisciplinary approach and appropriate treatment that involves, among others, ophthalmologists, physiotherapists, orthopedic surgeons, cardiologists, and cardiac surgeons. In this review, we focus on treatment of aortic disease. Improved diagnosis and better medical and surgical treatment yielded a substantial increase of more than 30 years in the life expectancy [113]. The cornerstone of adequate management of MFS is a correct and timely diagnosis, necessitating the adequate education of all involved health care workers and the correct echocardiographic assessment of the aortic dimensions by (pediatric) cardiologists according to established guidelines [114]. CT or MRI can be used in case of insufficient visualization of the ascending aorta by echocardiography. Once the diagnosis is confirmed, an initial reassessment of the aortic dimension is recommended after 6 months to determine evolutionary changes. Further follow-up is guided by the diameter, evolution, underlying diagnosis, and family history. Stable diameters <45 mm in MFS patients without a family history for dissection require yearly follow-up. Bi-annual controls are recommended in case of diameters exceeding 45 mm, increased growth rate (>2 mm/year), and a strong family history for aortic dissection. Despite rigorous follow-up and adherence to the guidelines, some MFS patients will still

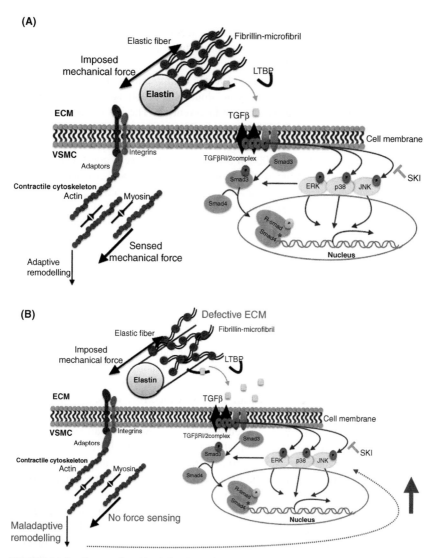

FIGURE 4A.3   Current hypothesis for the pathogenetic mechanism of Marfan syndrome. (A) In healthy individuals, cells sense and regulate the extracellular matrix via integrins/focal adhesion complexes and the contractile cytoskeleton. (B) Mutations in fibrillin-1 can affect the mechanical properties of the microfibrils leading to an altered mechanotransduction signal and initiation of cellular response mechanisms, including increased TGF-β signaling.

succumb to aortic dissection, indicating the need for improved risk stratification. Ongoing studies are identifying potential circulating biomarkers, including fibrillin-1 fragments, which may be used in conjunction with imaging data to better identify patients at risk for aortic dissection [115]. As mentioned above, other causes of sudden cardiac death, including arrhythmia, heart failure, and valvular heart disease, need to be taken into account; regular screenings with 24 h ambulatory ECG, NT-proBNP, and echocardiography to detect problems in an early stage are recommended.

The ultimate goal for treatment of aortic complications in MFS is to avoid aortic dissection or rupture. It is beyond doubt that prophylactic aortic root surgery is the most successful preventive intervention. The initial procedure was developed by Hugh Bentall and Antony de Bono and consisted of replacement by a composite graft with a mechanical aortic valve [116]. Results of this procedure in a large series of >600 MFS patients showed very good results when applied in an elective setting, with a mortality rate as low as 1.5%, as compared to a much higher rate of 11.7% in the setting of acute dissection [117]. Refinements of surgical techniques, mainly with the introduction of valve-sparing procedures, have led to excellent short- and middle-term results and have obviated the need for chronic use of anticoagulants [118]. Surgery is routinely recommended for MFS patients when the diameter of the aortic root reaches 50 mm. A reduced threshold of 45 mm is warranted in cases of rapid growth (>0.2–0.5 cm/year), desire for pregnancy, a family history for aortic dissection, and/or significant aortic valve regurgitation [72,119]. Although these conventional criteria have definitely proved their benefit, further refinement seems appropriate to select those cases at risk with lower diameters, on the one hand, and cases that may still be "safe" at higher diameters, on the other hand. Additional criteria that could be considered include (1) aortic biomechanics, (2) expected normal aortic dimensions, (3) aortic geometry and shape, and (4) specific biomarkers, such as fibrillin fragments or serum TGF-β. A nice review assessing some of these aspects has recently been published [120].

Because aortic surgery is inherently associated with a small risk for complications and a prolonged rehabilitation period, attempts have been made to at least postpone the need for surgery. Ultimately, arresting aortic growth and, conceptually, interfering with the underlying genetic defect are deemed to be the ideal solutions—which are, alas, not yet feasible.

Slowing down aortic root growth may be achieved through reducing hemodynamic stress on the proximal aorta. The first report on the use of beta-blockers in patients with MFS dates from 1971, indicating that reduction in the rate of increase in aortic pressure over time (dP/dt) was more effective than could be explained by reduction of blood pressure alone [121]. Subsequent small studies with beta-blockers in turkeys that are prone to aortic dissection and in uncontrolled studies of patients with MFS had varying results [122,123].

No fewer than 1583 MFS patients have been included in at least 19 different trials since the late 1970s [57,122,124–140]. However, only one trial had a controlled randomized design. This study, which showed a significant reduction in aortic root growth with propranolol, had a placebo [126]. Many other trials have confirmed the effect of beta-blockers on aortic growth, and in some trials, beneficial effects on elastic properties of the aorta have been reported [127].

This information notwithstanding, the beneficial effect of beta-blockers has not been consistent in all studies. Differences in the populations studied, in the drug types and dosages, and in the study designs have rendered the interpretation and comparison of these trials particularly challenging.

Another aspect that remains elusive despite numerous trials is the optimal timing of initiation of treatment: some advocate starting treatment as soon as the diagnosis is made, while others suggest waiting until some dilatation has occurred. Recently, arguments in favor of early treatment initiation were provided in a small, nonrandomized study as well as in a large randomized trial conducted by the Pediatric Heart Network [137,141]. Alternatives for beta-blockers have been suggested and studied in small series.

Calcium channel blockers have been investigated in a study with only six patients receiving calcium channel blockers, [128] which showing similar results as beta-blockers. ACE inhibitors (ACEIs) were postulated to have therapeutic value for patients with MFS because of their potential ability to block apoptotic pathways through inhibition of the AT2 receptor. An open-label, nonrandomized trial of ACEIs versus beta-blockers in patients with MFS showed apparent therapeutic value for ACEIs in terms of reduced aortic stiffness and smaller increases in aortic root diameter [130]. Another randomized trial of perindopril versus placebo in MFS revealed improved biomechanical properties of the aorta, slower aortic root growth, and lower levels of circulating TGF-β [133].

A seemingly major breakthrough in the search for improved medical treatment in MFS patients was achieved with the documentation of the involvement of the TGF-β pathway in the process of aneurysm formation. This led to the insight that interference with TGF-β signaling may have a beneficial impact on aortic growth. Losartan, an angiotensin II receptor blocker with known TGF-β-inhibiting potential, was tested in a mouse model for MFS that showed normalization of aortic root growth and restoration of elastic fiber fragmentation in the losartan-treated mice but not in the mice treated with propranolol [142]. Nonrandomized studies in humans showed similar results. A significantly decreased growth rate was observed in a small cohort study of 18 severely affected children with MFS who were treated with losartan on top of beta-blocker treatment [134] as well as in a small prospective study of 22 young MFS patients mainly treated with losartan in monotherapy [137]. A retrospective study comparing beta-blockers to valsartan showed an equal decline in growth rate in both groups [138]. Following the trial with MFS mice and the small pilot study in humans, at least 10 randomized trials ensued, recruiting >2000 patients in total, spanning all age ranges. The first published results in an open-label trial in a pediatric age group as well as in an open-label study in adult MFS patients showed a beneficial effect of the combination of beta-blockers with losartan [136,139]. But these observations could not be confirmed in a large double-blind randomized study, questioning the synergistic effect of both drugs [143]. Very recently, the results of a large randomized trial in a young (1 months–25 years) MFS cohort, conducted by the Pediatric Heart Network and comparing atenolol and losartan, were published and showed no difference in the aortic root growth rate between both groups [80,140,141]. The recommendations put forward in an editorial on the latter study were to continue considering beta-blockers as primary medical therapy for aortic protection in patients with MFS. Losartan is considered to be a reasonable alternative, especially in patients who cannot tolerate beta-blockers [144]. These results also indicate that the exact contribution of TGF-β in the process of aneurysm formation needs to be carefully reassessed.

Recently, a small substudy of the Compare trial indicated that patients harboring *FBN1* mutations leading to haploinsufficiency showed a better response to losartan than patients with mutations with a dominant negative effect [145].

To obtain a better insight into the effect of treatment in these and other specific subgroups, a large metaanalysis is currently being undertaken by Alex Pitcher and colleagues of the Marfan Treatment Trialist Collaboration. More than 2400 MFS patients will be included in this analysis, thus significantly increasing its statistical power [146].

6 PREGNANCY WITH MARFAN SYNDROME

The inherent hemodynamic and hormonal changes related to pregnancy impose extra stress on the fragile aorta in women with MFS and put them at an increased risk for aortic dissection during pregnancy as well as in the months following delivery. Reported data on the increased risk for dissection during pregnancy in general need to be interpreted with caution, because these are largely based on retrospective reports. The coincidental co-occurrence of pregnancy and dissection in young women may blur these reported figures [147]. To identify whether the association of pregnancy and acute type A dissection could be an artifact of selective reporting, Markus Thalmann et al. conducted an epidemiologic study, which failed to find a correlation [148].

In MFS patients in general, the risk for aortic dissection is estimated based on the aortic root diameter, and guidelines for prophylactic surgery are largely based on the measurement of aortic root dimensions, as discussed in detail above. Specific cut-off values for aortic root diameter have been suggested to estimate the risk during pregnancy, but caution is warranted, because dissection may also occur without significant dilatation [149]. The first review of pregnancy-associated cardiovascular risk in patients with MFS dates from 1981 [150]. A threshold of 40 mm for the aortic root diameter measured before pregnancy as a cut-off for categorizing women as having a low or increased risk of dissection (1 vs 10%) was proposed in this review. A prospective study comparing outcomes in women with aortic root diameters below and above 40 mm confirmed the beneficial outcome with regard to growth in the group with a diameter <40 mm [151]. Even though the threshold of 40 mm has been widely debated in the literature, [152–156] the current European guidelines for the management of cardiovascular diseases during pregnancy still adopt this measurement [157]. In the guidelines, maternal cardiovascular risk is classified as risk WHO II, III, or

IV, respectively, depending on whether there is no aortic dilation before pregnancy, the aortic dilation is between 40 and 45 mm, or the aortic dilation is >45 mm. In the latter situation, prophylactic aortic root surgery is recommended, although the outcome in patients with previous surgery is largely unknown, with adverse outcomes reported in a few cases [158]. Pregnancy in patients with previous aortic dissection should be discouraged. All MFS patients should undergo an extensive aortic imaging study (CT or MRI) prior to pregnancy.

In addition to the risk of dissection during or shortly after pregnancy, there is concern over the long-term effect of pregnancy in aortic root growth in patients with MFS. A prospective study indicated that aortic root growth rate increased during pregnancy and did not return to normal afterward. The prevalence of aortic dissection and elective aortic surgery during long-term follow-up was higher in those women who had a prior pregnancy [159].

Fetal outcome is another important factor to take into account. Higher rates of spontaneous abortion in MFS patients have been reported, as have associations between low birth weight and beta-blocker use [160,161].

Preconception counseling in MFS women is a class I recommendation in the European Society of Cardiology guidelines on management of pregnancy [157]. Genetic counseling should also address males and their 50% transmission risk. Information should be provided regarding reproductive options, including prenatal- and preimplantation diagnostics, donor insemination, and adoption. A prerequisite for prenatal and preimplantation diagnostics is that the underlying *FBN1* mutation be identified. In a French survey among patients and physicians, a majority of MFS patients seemed in favor of prenatal/preimplantation diagnostics. Large variation was noted regarding the physicians' standpoint, which was at least in part influenced by their experience with the disease [162].

Several recent prospective studies have reported that with detailed cardiovascular assessment prior to pregnancy, adequate preconception counseling, and strict medical follow-up and treatment with beta-blockers, pregnancy may be safe in patients with MFS. It appears that the most important factor influencing the outcome is knowledge of the diagnosis before pregnancy [159,163].

The many unanswered questions related to pregnancy in MFS emphasize the importance of large registries, such as the ROPAC registry, which will definitely help elucidate some of these issues and define clearer guidelines [164].

REFERENCES

[1] Judge DP, Dietz HC. Marfan's syndrome. Lancet 2005;366(9501):1965–76.

[2] Pyeritz RE, McKusick VA. The Marfan syndrome: diagnosis and management. New Engl J Med 1979;300(14):772–7.

[3] Von Kodolitsch Y, De Backer J, Schuler H, Bannas P, Behzadi C, Bernhardt AM, et al. Perspectives on the revised Ghent criteria for the diagnosis of Marfan syndrome. Appl Clin Genet 2015;8:137–55.

[4] Yang R-Q, Jabbari J, Cheng X-S, Jabbari R, Nielsen JB, Risgaard B, et al. New population-based exome data question the pathogenicity of some genetic variants previously associated with Marfan syndrome. BMC Genet 2014;15:74.

[5] McKusick VA. Heritable disorders of connective tissue. St Louis: Mosby; 1956.

[6] Beighton P, de Paepe A, Danks D, Finidori G, Gedde-Dahl T, Goodman R, et al. International nosology of heritable disorders of connective tissue, Berlin, 1986. Am J Med Genet 1988;29(3):581–94.

[7] de Paepe A, Devereux RB, Dietz HC, Hennekam RC, Pyeritz RE. Revised diagnostic criteria for the Marfan syndrome. Am J Med Genet 1996;62(4):417–26.

[8] Loeys BL, Dietz HC, Braverman AC, Callewaert BL, De Backer J, Devereux RB, et al. The revised Ghent nosology for the Marfan syndrome. J Med Genet 2010;47(7):476–85.

[9] Devereux RB, de Simone G, Arnett DK, Best LG, Boerwinkle E, Howard BV, et al. Normal limits in relation to age, body size and gender of two-dimensional echocardiographic aortic root dimensions in persons ≥15 years of age. Am J Cardiol 2012;110(8):1189–94.

[10] Campens L, Demulier L, De Groote K, Vandekerckhove K, De Wolf D, Roman MJ, et al. Reference values for echocardiographic assessment of the diameter of the aortic root and ascending aorta spanning all age categories. Am J Cardiol 2014;114(6):914–20.

[11] Pyeritz R. Marfan syndrome and related disorders. Emery and Rimoin's Essential Medical Genetics. Elsevier; 2013. p. 567–74.

[12] Rybczynski M, Treede H, Sheikhzadeh S, Groene EF, Bernhardt AMJ, Hillebrand M, et al. Predictors of outcome of mitral valve prolapse in patients with the Marfan syndrome. Am J Cardiol 2011;107(2):268–74.

[13] Dietz HC, Cutting GR, Pyeritz RE, Maslen CL, Sakai LY, Corson GM, et al. Marfan syndrome caused by a recurrent de novo missense mutation in the fibrillin gene. Nature 1991;352(6333):337–9.

[14] Collod-Beroud G, Le Bourdelles S, Adès L, Ala-Kokko L, Booms P, Boxer M, et al. Update of the UMD-FBN1 mutation database and creation of an FBN1 polymorphism database. Hum Mutat 2003;22(3):199–208.

[15] Pyeritz RE. The Marfan syndrome. Annu Rev Med 2000;51:481–510.

[16] Collod-Beroud G, Lackmy-Port-Lys M, Jondeau G, Mathieu M, Maingourd Y, Coulon M, et al. Demonstration of the recurrence of Marfan-like skeletal and cardiovascular manifestations due to germline mosaicism for an FBN1 mutation. Am J Hum Genet 1999;65(3): 917–21.

[17] Tekin M, Cengiz FB, Ayberkin E, Kendirli T, Fitoz S, Tutar E, et al. Familial neonatal Marfan syndrome due to parental mosaicism of a missense mutation in the FBN1 gene. Am J Med Genet A 2007;143A(8):875–80.

[18] Murdoch JL, Walker BA, McKusick VA. Parental age effects on the occurrence of new mutations for the Marfan syndrome. Ann Hum Genet 1972;35(3):331–6.

[19] Schrijver I, Liu W, Brenn T, Furthmayr H, Francke U. Cysteine substitutions in epidermal growth factor-like domains of fibrillin-1: distinct effects on biochemical and clinical phenotypes. Am J Hum Genet 1999;65(4):1007–20.

[20] Schrijver I, Liu W, Odom R, Brenn T, Oefner P, Furthmayr H, et al. Premature termination mutations in FBN1: distinct effects on differential allelic expression and on protein and clinical phenotypes. Am J Hum Genet 2002;71(2):223–37.

[21] Comeglio P, Evans AL, Brice G, Cooling RJ, Child AH. Identification of FBN1 gene mutations in patients with ectopia lentis and marfanoid habitus. Brit J Ophthalmol 2002;86(12):1359–62.

[22] Comeglio P, Johnson P, Arno G, Brice G, Evans A, Aragon-Martin J, et al. The importance of mutation detection in Marfan syndrome and Marfan-related disorders: report of 193 FBN1 mutations. Hum Mutat 2007;28(9):928.

[23] Faivre L, Masurel-Paulet A, Collod-Beroud G, Callewaert BL, Child AH, Stheneur C, et al. Clinical and molecular study of 320 children with Marfan syndrome and related type I fibrillinopathies in a series of 1009 probands with pathogenic FBN1 mutations. Pediatrics 2009;123(1):391–8.

[24] Faivre L, Collod-Beroud G, Loeys BL, Child A, Binquet C, Gautier E, et al. Effect of mutation type and location on clinical outcome in 1,013 probands with Marfan syndrome or related phenotypes and FBN1 mutations: an international study. Am J Hum Genet 2007;81(3):454–66.

[25] Loeys B, Nuytinck L, Delvaux I, De Bie S, de Paepe A. Genotype and phenotype analysis of 171 patients referred for molecular study of the fibrillin-1 gene FBN1 because of suspected Marfan syndrome. Arch Intern Med 2001;161(20):2447–54.

[26] Palz M, Tiecke F, Booms P, Göldner B, Rosenberg T, Fuchs J, et al. Clustering of mutations associated with mild Marfan-like phenotypes in the 3′ region of FBN1 suggests a potential genotype-phenotype correlation. Am J Med Genet 2000;91(3):212–21.

[27] Baudhuin LM, Kotzer KE, Lagerstedt SA. Increased frequency of FBN1 truncating and splicing variants in Marfan syndrome patients with aortic events. Genet Med 2015;17(3):177–87.

[28] de Backer J, Loeys B, Leroy B, Coucke P, Dietz H, de Paepe A. Utility of molecular analyses in the exploration of extreme intrafamilial variability in the Marfan syndrome. Clin Genet 2007;72(3):188–98.

[29] Wang M, Kishnani P, Decker-Phillips M, Kahler SG, Chen YT, Godfrey M. Double mutant fibrillin-1 (FBN1) allele in a patient with neonatal Marfan syndrome. J Med Genet 1996;33(9):760–3.

[30] Aubart M, Gross M-S, Hanna N, Zabot M-T, Sznajder M, Detaint D, et al. The clinical presentation of Marfan syndrome is modulated by expression of wildtype FBN1 allele. Hum Mol Genet 2015;24(10):2764–70.

[31] Hutchinson S, Furger A, Halliday D, Judge DP, Jefferson A, Dietz HC, et al. Allelic variation in normal human FBN1 expression in a family with Marfan syndrome: a potential modifier of phenotype? Hum Mol Genet 2003;12(18):2269–76.

[32] Giusti B, Marcucci R, Lapini I, Sestini I, Lenti M, Yacoub M, et al. Role of hyperhomocysteinemia in aortic disease. Cell Mol Biol (Noisy-le-grand) 2004;50(8):945–52.

[33] Faivre L, Collod-Beroud G, Callewaert B, Child A, Loeys BL, Binquet C, et al. Pathogenic FBN1 mutations in 146 adults not meeting clinical diagnostic criteria for Marfan syndrome: further delineation of type 1 fibrillinopathies and focus on patients with an isolated major criterion. Am J Med Genet A 2009;149A(5):854–60.

[34] Dietz HC, Pyeritz RE. Mutations in the human gene for fibrillin-1 (FBN1) in the Marfan syndrome and related disorders. Hum Mol Genet 1995;4:1799–809.

[35] Adès LC, Sreetharan D, Onikul E, Stockton V, Watson KC, Holman KJ. Segregation of a novel FBN1 gene mutation, G1796E, with kyphoscoliosis and radiographic evidence of vertebral dysplasia in three generations. Am J Med Genet 2002;109(4):261–70.

[36] Francke U, Berg MA, Tynan K, Brenn T, Liu W, Aoyama T, et al. A Gly1127Ser mutation in an EGF-like domain of the fibrillin-1 gene is a risk factor for ascending aortic aneurysm and dissection. Am J Hum Genet 1995;56(6):1287–96.

[37] Campens L, Callewaert B, Muiño Mosquera L, Renard M, Symoens S, de Paepe A, et al. Gene panel sequencing in heritable thoracic aortic disorders and related entities—results of comprehensive testing in a cohort of 264 patients. Orphanet J Rare Dis 2015;10:9.

[38] Chandra A, Patel D, Aragon-Martin JA, Pinard A, Collod-Beroud G, Comeglio P, et al. The revised ghent nosology; reclassifying isolated ectopia lentis. Clin Genet 2015;87(3):284–7.

[39] Le Goff C, Mahaut C, Wang LW, Allali S, Abhyankar A, Jensen S, et al. Mutations in the TGFβ binding-protein-like domain 5 of FBN1 are responsible for acromicric and geleophysic dysplasias. Am J Hum Genet 2011;89(1):7–14.

[40] Faivre L, Gorlin RJ, Wirtz MK, Godfrey M, Dagoneau N, Samples JR, et al. In frame fibrillin-1 gene deletion in autosomal dominant Weill-Marchesani syndrome. J Med Genet 2003;40(1):34–6.

[41] Loeys BL, Gerber EE, Riegert-Johnson D, Iqbal S, Whiteman P, McConnell V, et al. Mutations in fibrillin-1 cause congenital scleroderma: stiff skin syndrome. Sci Transl Med 2010;2(23):23ra20.

[42] Kosaki K, Takahashi D, Udaka T, Kosaki R, Matsumoto M, Ibe S, et al. Molecular pathology of Shprintzen-Goldberg syndrome. Am J Med Genet A 2006;140(1):104–8. author reply 109–110.

[43] Das BB, Taylor AL, Yetman AT. Left ventricular diastolic dysfunction in children and young adults with Marfan syndrome. Pediatr Cardiol 2006;27(2):256–8.

[44] Savolainen A, Nisula L, Keto P, Hekali P, Viitasalo M, Kaitila I, et al. Left ventricular function in children with the Marfan syndrome. Eur Heart J 1994;15(5):625–30.

[45] Kiotsekoglou A, Moggridge JC, Bijnens BH, Kapetanakis V, Alpendurada F, Mullen MJ, et al. Biventricular and atrial diastolic function assessment using conventional echocardiography and tissue-Doppler imaging in adults with Marfan syndrome. Eur J Echocardiogr 2009;10(8):947–55.

[46] Alpendurada F, Wong J, Kiotsekoglou A, Banya W, Child A, Prasad SK, et al. Evidence for Marfan cardiomyopathy. Eur J Heart Fail 2010;12(10):1085–91.

[47] de Backer JF, Devos D, Segers P, Matthys D, François K, Gillebert TC, et al. Primary impairment of left ventricular function in Marfan syndrome. Int J Cardiol 2006;112(3):353–8.

[48] Rybczynski M, Koschyk DH, Aydin MA, Robinson PN, Brinken T, Franzen O, et al. Tissue Doppler imaging identifies myocardial dysfunction in adults with Marfan syndrome. Clin Cardiol 2007;30(1):19–24.

[49] Schoenhoff FS, Jungi S, Czerny M, Roost E, Reineke D, Mátyás G, et al. Acute aortic dissection determines the fate of initially untreated aortic segments in Marfan syndrome. Circulation 2013;127(15):1569–75.

[50] McKusick VA. The cardiovascular aspects of Marfan's syndrome: a heritable disorder of connective tissue. Circulation 1955;11(3):321–42.

[51] Sherif HMF. Heterogeneity in the segmental development of the aortic tree: impact on management of genetically triggered aortic aneurysms. Aorta (Stamford) 2014;2(5):186–95.

[52] Hungerford JE, Little CD. Developmental biology of the vascular smooth muscle cell: building a multilayered vessel wall. J Vasc Res 1999;36(1):2–27.

[53] Teixido-Tura G, Redheuil A, Rodríguez-Palomares J, Gutiérrez L, Sánchez V, Forteza A, et al. Aortic biomechanics by magnetic resonance: early markers of aortic disease in Marfan syndrome regardless of aortic dilatation? Int J Cardiol 2014;171(1):56–61.

[54] Lindsay ME, Dietz HC. Lessons on the pathogenesis of aneurysm from heritable conditions. Nature 2011;473(7347):308–16.

[55] Apter JT. Correlation of visco-elastic properties with microscopic structure of large arteries. IV. Thermal responses of collagen, elastin, smooth muscle, and intact arteries. Circ Res 1967;21(6):901–18.

[56] Yetman AT, Bornemeier RA, McCrindle BW. Long-term outcome in patients with Marfan syndrome: is aortic dissection the only cause of sudden death? J Am Coll Cardiol 2003;41(2):329–32.

[57] Reed CM, Fox ME, Alpert BS. Aortic biomechanical properties in pediatric patients with the Marfan syndrome, and the effects of atenolol. Am J Cardiol 1993;71(7):606–8.

[58] Hoffmann BA, Rybczynski M, Rostock T, Servatius H, Drewitz I, Steven D, et al. Prospective risk stratification of sudden cardiac death in Marfan's syndrome. Int J Cardiol 2013;167(6):2539–45.

[59] Roman MJ, Rosen SE, Kramer-Fox R, Devereux RB. Prognostic significance of the pattern of aortic root dilation in the Marfan syndrome. J Am Coll Cardiol 1993;22(5):1470–6.

[60] Chen S, Fagan LF, Nouri S, Donahoe JL. Ventricular dysrhythmias in children with Marfan's syndrome. Am J Dis Child 1985;139(3):273–6.

[61] Pyeritz RE. The Marfan syndrome. Am Fam Physician 1986;34(6):83–94.

[62] Keane MG, Pyeritz RE. Medical management of Marfan syndrome. Circulation 2008;117(21):2802–13.

[63] Lundby R, Rand-Hendriksen S, Hald JK, Pripp AH, Smith H-J. The pulmonary artery in patients with Marfan syndrome: a cross-sectional study. Genet Med 2012;14(11):922–7.

[64] de Backer J, Loeys B, Devos D, Dietz H, de Sutter J, de Paepe A. A critical analysis of minor cardiovascular criteria in the diagnostic evaluation of patients with Marfan syndrome. Genet Med 2006;8(7):401–8.

[65] Sheikhzadeh S, De Backer J, Gorgan N, Rybczynski M, Hillebrand M, Schüler H, et al. The main pulmonary artery in adults: a controlled multicenter study with assessment of echocardiographic reference values, and the frequency of dilatation and aneurysm in Marfan syndrome. Orphanet J Rare Dis 2014;9:203.

[66] Finkbohner R, Johnston D, Crawford ES, Coselli J, Milewicz DM. Marfan syndrome. Long-term survival and complications after aortic aneurysm repair. Circulation 1995;91(3): 728–33.

[67] Engelfriet PM, Boersma E, Tijssen JG, Bouma BJ, Mulder BJ. Beyond the root: dilatation of the distal aorta in Marfan's syndrome. Heart 2006;92(9):1238–43.

[68] van Ooijen B. Marfan's syndrome and isolated aneurysm of the abdominal aorta. Brit Heart J 1988;59(1):81–4.

[69] Pruzinsky MS, Katz NM, Green CE, Satler LF. Isolated descending thoracic aortic aneurysm in Marfan's syndrome. Am J Cardiol 1988;61(13):1159–60.

[70] Mimoun L, Detaint D, Hamroun D, Arnoult F, Delorme G, Gautier M, et al. Dissection in Marfan syndrome: the importance of the descending aorta. Eur Heart J 2011;32(4):443–9.

[71] Lemaire SA, la Cruz de KI, Coselli JS. The thoracoabdominal aorta in Marfan syndrome. London: Springer London; 2014. p. 423–434.

[72] Hiratzka LF, Bakris GL, Beckman JA, Bersin RM, Carr VF, Casey DE, et al. 2010 ACCF/ AHA/AATS/ACR/ASA/SCA/SCAI/SIR/STS/SVM guidelines for the diagnosis and management of patients with Thoracic Aortic Disease: a report of the American College of Cardiology Foundation/American Heart Association Task Force on Practice Guidelines, American Association for Thoracic Surgery, American College of Radiology, American Stroke Association, Society of Cardiovascular Anesthesiologists, Society for Cardiovascular Angiography and Interventions, Society of Interventional Radiology, Society of Thoracic Surgeons, and Society for Vascular Medicine. Circulation. 2010. pp. e266–e369.

[73] Groenink M, de Roos A, Mulder BJ, Verbeeten B, Timmermans J, Zwinderman AH, et al. Biophysical properties of the normal-sized aorta in patients with Marfan syndrome: evaluation with MR flow mapping. Radiology 2001;219(2):535–40.

[74] Nollen GJ, Meijboom LJ, Groenink M, Timmermans J, Barentsz JO, Merchant N, et al. Comparison of aortic elasticity in patients with the Marfan syndrome with and without aortic root replacement. Am J Cardiol 2003;91(5):637–40.

[75] Nollen GJ, Groenink M, Tijssen JGP, Van Der Wall EE, Mulder BJM. Aortic stiffness and diameter predict progressive aortic dilatation in patients with Marfan syndrome. Eur Heart J 2004;25(13):1146–52.

[76] Mortensen K, Aydin MA, Rybczynski M, Baulmann J, Schahidi NA, Kean G, et al. Augmentation index relates to progression of aortic disease in adults with Marfan syndrome. Am J Hypertens 2009;22(9):971–9.

[77] Segers P, de Backer J, Devos D, Rabben SI, Gillebert TC, Van Bortel LM, et al. Aortic reflection coefficients and their association with global indexes of wave reflection in healthy controls and patients with Marfan's syndrome. Am J Physiol-Heart Circ Physiol 2006;290(6):H2385–92.

[78] Rybczynski M, Mir TS, Sheikhzadeh S, Bernhardt AMJ, Schad C, Treede H, et al. Frequency and age-related course of mitral valve dysfunction in the Marfan syndrome. Am J Cardiol 2010;106(7):1048–53.

[79] Selamet Tierney ES, Levine JC, Chen S, Bradley TJ, Pearson GD, Colan SD, et al. Echocardiographic methods, quality review, and measurement accuracy in a randomized multicenter clinical trial of Marfan syndrome. J Am Soc Echocardiog 2013;26(6):657–66.

[80] Lacro RV, Guey LT, Dietz HC, Pearson GD, Yetman AT, Gelb BD, et al. Characteristics of children and young adults with Marfan syndrome and aortic root dilation in a randomized trial comparing atenolol and losartan therapy. Am Heart J 2013;165(5):828–35.

[81] Detaint D, Faivre L, Collod-Beroud G, Child AH, Loeys BL, Binquet C, et al. Cardiovascular manifestations in men and women carrying a FBN1 mutation. Eur Heart J 2010;31(18):2223–9.

[82] Pyeritz RE, Wappel MA. Mitral valve dysfunction in the Marfan syndrome. Clinical and echocardiographic study of prevalence and natural history. Am J Med 1983;74(5):797–807.

[83] Knosalla C, Weng Y-G, Hammerschmidt R, Pasic M, Schmitt-Knosalla I, Grauhan O, et al. Orthotopic heart transplantation in patients with Marfan syndrome. Ann Thorac Surg 2007;83(5):1691–5.

[84] Chatrath R, Beauchesne LM, Connolly HM, Michels VV, Driscoll DJ. Left ventricular function in the Marfan syndrome without significant valvular regurgitation. Am J Cardiol 2003;91(7):914–6.

[85] Campens L, Renard M, Trachet B, Segers P, Muino Mosquera L, De Sutter J, et al. Intrinsic cardiomyopathy in Marfan syndrome: results from in-vivo and ex-vivo studies of Fbn1C1039G/+ model and longitudinal findings in humans. Pediatr Res 2015;78:256–63.

[86] Aalberts JJJ, van Tintelen JP, Meijboom LJ, Polko A, Jongbloed JDH, van der Wal H, et al. Relation between genotype and left-ventricular dilatation in patients with Marfan syndrome. Gene 2014;534(1):40–3.

[87] Cook JR, Carta L, Bénard L, Chemaly ER, Chiu E, Rao SK, et al. Abnormal muscle mechanosignaling triggers cardiomyopathy in mice with Marfan syndrome. J Clin Invest 2014;124(3):1329–39.

[88] Attias D, Stheneur C, Roy C, Collod-Beroud G, Detaint D, Faivre L, et al. Comparison of clinical presentations and outcomes between patients with TGFBR2 and FBN1 mutations in Marfan syndrome and related disorders. Circulation 2009;120(25):2541–9.

[89] Grahame R, Pyeritz RE. The Marfan syndrome: joint and skin manifestations are prevalent and correlated. Brit J Rheumatol 1995;34(2):126–31.

[90] Grover M, Brunetti-Pierri N, Belmont J, Phan K, Tran A, Shypailo RJ, et al. Assessment of bone mineral status in children with Marfan syndrome. Am J Med Genet A 2012;158A(9):2221–4.

[91] Haine E, Salles J-P, van Kien PK, Conte-Auriol F, Gennero I, Plancke A, et al. Muscle and bone impairment in children with Marfan syndrome: correlation with age and FBN1 genotype. J Bone Miner Res 2015;30(8):1369–76.

[92] Maumenee IH. The eye in the Marfan syndrome. T Am Ophthal Soc 1981;79:684–733.

[93] Drolsum L, Rand-Hendriksen S, Paus B, Geiran OR, Semb SO. Ocular findings in 87 adults with Ghent-1 verified Marfan syndrome. Acta Ophthalmol 2015;93(1):46–53.

[94] Streeten EA, Murphy EA, Pyeritz RE. Pulmonary function in the Marfan syndrome. Chest 1987;91(3):408–12.

[95] Wood JR, Bellamy D, Child AH, Citron KM. Pulmonary disease in patients with Marfan syndrome. Thorax 1984;39(10):780–4.

[96] Hall JR, Pyeritz RE, Dudgeon DL, Haller JA JJr. Pneumothorax in the Marfan syndrome: prevalence and therapy. Ann Thorac Surg 1984;37(6):500–4.

[97] Neptune ER, Frischmeyer PA, Arking DE, Myers L, Bunton TE, Gayraud B, et al. Dysregulation of TGF-beta activation contributes to pathogenesis in Marfan syndrome. Nat Genet 2003;33(3):407–11.

[98] Rybczynski M, Koschyk D, Karmeier A, Gessler N, Sheikhzadeh S, Bernhardt AMJ, et al. Frequency of sleep apnea in adults with the Marfan syndrome. Am J Cardiol 2010;105(12):1836–41.

[99] Kohler M, Pitcher A, Blair E, Risby P, Senn O, Forfar C, et al. The impact of obstructive sleep apnea on aortic disease in Marfan's syndrome. Respiration 2013;86(1):39–44.

[100] Sakai LY, Keene DR, Engvall E. Fibrillin, a new 350-kD glycoprotein, is a component of extracellular microfibrils. J Cell Biol 1986;103(6 Pt 1):2499–509.

[101] Hollister DW, Godfrey M, Sakai LY, Pyeritz RE. Immunohistologic abnormalities of the microfibrillar-fiber system in the Marfan syndrome. New Engl J Med 1990;323(3): 152–9.

[102] Lee B, Godfrey M, Vitale E, Hori H, Mattei MG, Sarfarazi M, et al. Linkage of Marfan syndrome and a phenotypically related disorder to two different fibrillin genes. Nature 1991;352(6333):330–4.

[103] Wilson DG, Bellamy MF, Ramsey MW, Goodfellow J, Brownlee M, Davies S, et al. Endothelial function in Marfan syndrome: selective impairment of flow-mediated vasodilation. Circulation 1999;99(7):909–15.

[104] Chung AWY, Au Yeung K, Cortes SF, Sandor GGS, Judge DP, Dietz HC, et al. Endothelial dysfunction and compromised eNOS/Akt signaling in the thoracic aorta during the progression of Marfan syndrome. Brit J Pharmacol 2007;150(8):1075–83.

[105] Isogai Z, Ono RN, Ushiro S, Keene DR, Chen Y, Mazzieri R, et al. Latent transforming growth factor beta-binding protein 1 interacts with fibrillin and is a microfibril-associated protein. J Biol Chem 2003;278(4):2750–7.

[106] Dallas SL, Miyazono K, Skerry TM, Mundy GR, Bonewald LF. Dual role for the latent transforming growth factor-beta binding protein in storage of latent TGF-beta in the extracellular matrix and as a structural matrix protein. J Cell Biol 1995;131(2):539–49.

[107] Dallas SL, Keene DR, Bruder SP, Saharinen J, Sakai LY, Mundy GR, et al. Role of the latent transforming growth factor beta binding protein 1 in fibrillin-containing microfibrils in bone cells in vitro and in vivo. J Bone Miner Res 2000;15(1):68–81.

[108] Saharinen J, Hyytiäinen M, Taipale J, Keski-Oja J. Latent transforming growth factor-beta binding proteins (LTBPs)—structural extracellular matrix proteins for targeting TGF-beta action. Cytokine Growth F R 1999;10(2):99–117.

[109] Charbonneau NL, Carlson EJ, Tufa S, Sengle G, Manalo EC, Carlberg VM, et al. In vivo studies of mutant fibrillin-1 microfibrils. J Biol Chem 2010;285(32):24943–55.

[110] Dietz HC. TGF-beta in the pathogenesis and prevention of disease: a matter of aneurysmic proportions. J Clin Invest 2010;120(2):403–6.

[111] Jeremy RW, Robertson E, Lu Y, Hambly BD. Perturbations of mechanotransduction and aneurysm formation in heritable aortopathies. Int J Cardiol 2013;169(1):7–16.

[112] Horiguchi M, Ota M, Rifkin DB. Matrix control of transforming growth factor-β function. J Biochem 2012;152(4):321–9.

[113] Silverman DI, Burton KJ, Gray J, Bosner MS, Kouchoukos NT, Roman MJ, et al. Life expectancy in the Marfan syndrome. Am J Cardiol 1995;75(2):157–60.

[114] Evangelista A, Flachskampf FA, Erbel R, Antonini-Canterin F, Vlachopoulos C, Rocchi G, et al. Echocardiography in aortic diseases: EAE recommendations for clinical practice. Eur J Echocardiogr 2010;11(8):645–58.

[115] Marshall LM, Carlson EJ, O'Malley J, Snyder CK, Charbonneau NL, Hayflick SJ, et al. Thoracic aortic aneurysm frequency and dissection are associated with fibrillin-1 fragment concentrations in circulation. Circ Res 2013;113(10):1159–68.

[116] Bentall H, De Bono A. A technique for complete replacement of the ascending aorta. Thorax 1968;23(4):338–9.

[117] Gott VL, Greene PS, Alejo DE, Cameron DE, Naftel DC, Miller DC, et al. Replacement of the aortic root in patients with Marfan's syndrome. New Engl J Med 1999;340(17):1307–13.

[118] Svensson LG, Blackstone EH, Alsalihi M, Batizy LH, Roselli EE, McCullough R, et al. Mid-term results of David reimplantation in patients with connective tissue disorder. Ann Thorac Surg 2013;95(2):555–62.

[119] Baumgartner H, Bonhoeffer P, De Groot NMS, de Haan F, Deanfield JE, Galie N, et al. ESC guidelines for the management of grown-up congenital heart disease (new version 2010). Eur Heart J 2010;31(23):2915–57.

[120] von Kodolitsch Y, Robinson PN, Berger J. When should surgery be performed in Marfan syndrome and other connective tissue disorders to protect against type A dissection? London: Springer London; 2014. p. 17–47.

[121] Halpern BL, Char F, Murdoch JL, Horton WB, McKusick VA. A prospectus on the prevention of aortic rupture in the Marfan syndrome with data on survivorship without treatment. Johns Hopkins Med J 1971;129(3):123–9.

[122] Ose L, McKusick VA. Prophylactic use of propranolol in the Marfan syndrome to prevent aortic dissection. Birth Def 1977;13(3C):163–9.

[123] Simpson CF, Boucek RJ, Noble NL. Influence of d-, l-, and dl-propranolol, and practolol on beta-amino-propionitrile-induced aortic ruptures of turkeys. Toxicol Appl Pharm 1976;38(1):169–75.

[124] Salim MA, Alpert BS, Ward JC, Pyeritz RE. Effect of beta-adrenergic blockade on aortic root rate of dilation in the Marfan syndrome. Am J Cardiol 1994;74(6):629–33.

[125] Tahernia AC. Cardiovascular anomalies in Marfan's syndrome: the role of echocardiography and beta-blockers. South Med J 1993;86(3):305–10.

[126] Shores J, Berger KR, Murphy EA, Pyeritz RE. Progression of aortic dilatation and the benefit of long-term beta-adrenergic blockade in Marfan's syndrome. New Engl J Med 1994;330(19):1335–41.

[127] Groenink M, de Roos A, Mulder BJ, Spaan JA, van der Wall EE. Changes in aortic distensibility and pulse wave velocity assessed with magnetic resonance imaging following beta-blocker therapy in the Marfan syndrome. Am J Cardiol 1998;82(2):203–8.

[128] Rossi-Foulkes R, Roman MJ, Rosen SE, Kramer-Fox R, Ehlers KH, O'Loughlin JE, et al. Phenotypic features and impact of beta blocker or calcium antagonist therapy on aortic lumen size in the Marfan syndrome. Am J Cardiol 1999;83(9):1364–8.

[129] Rios AS, Silber EN, Bavishi N, Varga P, Burton BK, Clark WA, et al. Effect of long-term beta-blockade on aortic root compliance in patients with Marfan syndrome. Am Heart J 1999;137(6):1057–61.

[130] Yetman AT, Bornemeier RA, McCrindle BW. Usefulness of enalapril versus propranolol or atenolol for prevention of aortic dilation in patients with the Marfan syndrome. Am J Cardiol 2005;95(9):1125–7.

[131] Ladouceur M, Fermanian C, Lupoglazoff J-M, Edouard T, Dulac Y, Acar P, et al. Effect of beta-blockade on ascending aortic dilatation in children with the Marfan syndrome. Am J Cardiol 2007;99(3):406–9.

[132] Selamet Tierney ES, Feingold B, Printz BF, Park SC, Graham D, Kleinman CS, et al. Beta-blocker therapy does not alter the rate of aortic root dilation in pediatric patients with Marfan syndrome. J Pediatr 2007;150(1):77–82.

[133] Ahimastos AA, Aggarwal A, D'Orsa KM, Formosa MF, White AJ, Savarirayan R, et al. Effect of perindopril on large artery stiffness and aortic root diameter in patients with Marfan syndrome: a randomized controlled trial. JAMA-J Am Med Assoc 2007;298(13):1539–47.

[134] Brooke BS, Habashi JP, Judge DP, Patel N, Loeys B, Dietz HC. Angiotensin II blockade and aortic-root dilation in Marfan's syndrome. New Engl J Med 2008;358(26):2787–95.

[135] Williams A, Kenny D, Wilson D, Fagenello G, Nelson M, Dunstan F, et al. Effects of atenolol, perindopril and verapamil on haemodynamic and vascular function in Marfan syndrome—a randomised, double-blind, crossover trial. Eur J Clin Invest 2012;42(8):891–9.

[136] Chiu H-H, Wu M-H, Wang J-K, Lu C-W, Chiu S-N, Chen C-A, et al. Losartan added to β-blockade therapy for aortic root dilation in Marfan syndrome: a randomized, open-label pilot study. Mayo Clin Proc 2013;88(3):271–6.

[137] Pees C, Laccone F, Hagl M, Debrauwer V, Moser E, Michel-Behnke I. Usefulness of losartan on the size of the ascending aorta in an unselected cohort of children, adolescents, and young adults with Marfan syndrome. Am J Cardiol 2013;112(9):1477–83.

[138] Mueller GC, Stierle L, Stark V, Steiner K, Kodolitsch von Y, Weil J, et al. Retrospective analysis of the effect of angiotensin II receptor blocker versus β-blocker on aortic root growth in paediatric patients with Marfan syndrome. Heart 2014;100(3):214–8.

[139] Groenink M, Hartog den AW, Franken R, Radonic T, de Waard V, Timmermans J, et al. Losartan reduces aortic dilatation rate in adults with Marfan syndrome: a randomized controlled trial. Eur Heart J 2013;34(45):3491–500.

[140] Lacro RV, Dietz HC, Wruck LM, Bradley TJ, Colan SD, Devereux RB, et al. Rationale and design of a randomized clinical trial of beta-blocker therapy (atenolol) versus angiotensin II receptor blocker therapy (losartan) in individuals with Marfan syndrome. Am Heart J 2007;154(4):624–31.

[141] Lacro RV, Dietz HC, Sleeper LA, Yetman AT, Bradley TJ, Colan SD, et al. Atenolol versus losartan in children and young adults with Marfan's syndrome. New Engl J Med 2014;371(22):2061–71.

[142] Habashi JP, Judge DP, Holm TM, Cohn RD, Loeys BL, Cooper TK, et al. Losartan, an AT1 antagonist, prevents aortic aneurysm in a mouse model of Marfan syndrome. Science 2006;312(5770):117–21.

[143] Jondeau G. Results of French Trial "Marfan Sartan." Paris.

[144] Bowen JM, Connolly HM. Of Marfan's syndrome, mice, and medications. New Engl J Med 2014;371(22):2127–8.

[145] Franken R, den Hartog AW, Radonic T, Micha D, Maugeri A, van Dijk FS, et al. Beneficial outcome of losartan therapy depends on type of FBN1 mutation in Marfan syndrome. Circ Cardiovasc Genet 2015;8(2):383–8.

[146] Pitcher A, Emberson J, Lacro RV, Sleeper LA, Stylianou M, Mahony L. Design and rationale of a prospective, collaborative meta-analysis of all randomized controlled trials of angiotensin receptor antagonists in Marfan syndrome, based on individual patient data: A report from the Marfan Treatment Trialists' Collaboration. Am Heart J 2015;169:605–12.

[147] Oskoui R, Lindsay J. Aortic dissection in women. Am J Cardiol 1994;73(11):821–3.

[148] Thalmann M, Sodeck GH, Domanovits H, Grassberger M, Loewe C, Grimm M, et al. Acute type A aortic dissection and pregnancy: a population-based study. Eur J Cardiothorac Surg 2011;39(6):e159–63.

[149] Lipscomb KJ, Smith JC, Clarke B, Donnai P, Harris R. Outcome of pregnancy in women with Marfan's syndrome. Brit J Obstet Gynaec 1997;104(2):201–6.

[150] Pyeritz RE. Maternal and fetal complications of pregnancy in the Marfan syndrome. Am J Med 1981;71(5):784–90.

[151] Meijboom LJ. Pregnancy and aortic root growth in the Marfan syndrome: a prospective study. Eur Heart J 2005;26(9):914–20.

[152] Elkayam U, Ostrzega E, Shotan A, Mehra A. Cardiovascular problems in pregnant women with the Marfan syndrome. Ann Intern Med 1995;123(2):117–22.

[153] Goland S, Elkayam U. Cardiovascular problems in pregnant women with Marfan syndrome. Circulation 2009;119(4):619–23.

[154] Lalchandani S, Wingfield M. Pregnancy in women with Marfan's syndrome. Eur J Obstet Gyn R B 2003;110(2):125–30.

[155] Lind J, Wallenburg HC. The Marfan syndrome and pregnancy: a retrospective study in a Dutch population. Eur J Obstet Gyn R B 2001;98(1):28–35.

[156] Pacini L, Digne F, Boumendil A, Muti C, Detaint D, Boileau C, et al. Maternal complication of pregnancy in Marfan syndrome. Int J Cardiol 2009;136(2):156–61.

[157] Regitz-Zagrosek V, Blomstrom Lundqvist C, Borghi C, European Society of Gynecology (ESG), Association for European Paediatric Cardiology (AEPC), German Society for Gender Medicine (DGesGM). et al. ESC Guidelines on the management of cardiovascular diseases during pregnancy: the Task Force on the Management of Cardiovascular Diseases during Pregnancy of the European Society of Cardiology (ESC). Eur Heart J 2011;32(24):3147–97.

[158] Mulder BJM, Meijboom LJ. Pregnancy and Marfan syndrome. J Am Coll Cardiol 2012;60(3):230–1.

[159] Donnelly RT, Pinto NM, Kocolas I, Yetman AT. The immediate and long-term impact of pregnancy on aortic growth rate and mortality in women with Marfan syndrome. J Am Coll Cardiol 2012;60(3):224–9.

[160] Ruys TPE, Maggioni A, Johnson MR, Sliwa K, Tavazzi L, Schwerzmann M, et al. Cardiac medication during pregnancy, data from the ROPAC. Int J Cardiol 2014;177(1):124–8.

[161] Rossiter JP, Repke JT, Morales AJ, Murphy EA, Pyeritz RE. A prospective longitudinal evaluation of pregnancy in the Marfan syndrome. Am J Obstet Gynecol 1995;173(5):1599–606.

[162] Coron F, Rousseau T, Jondeau G, Gautier E, Binquet C, Gouya L, et al. What do French patients and geneticists think about prenatal and preimplantation diagnoses in Marfan syndrome? Prenatal Diag 2012;32(13):1318–23.

[163] Omnes S, Jondeau G, Detaint D, Dumont A, Yazbeck C, Guglielminotti J, et al. Pregnancy outcomes among women with Marfan syndrome. Int J Gynecol Obstet 2013;122(3):219–23.

[164] Roos-Hesselink JW, Ruys TPE, Stein JI, Thilen U, Webb GD, Niwa K, et al. Outcome of pregnancy in patients with structural or ischaemic heart disease: results of a registry of the European Society of Cardiology. Eur Heart J 2013;34(9):657–65.

Chapter 4b

Loeys-Dietz Syndrome

B.L. Loeys, MD, PhD

1 INITIAL DESCRIPTION

In 2005, Bart Loeys and Harry Dietz described a new syndromic aneurysmal entity. Its patients were typically characterized by a clinical triad that included hypertelorism (widely spaced eyes), cleft palate, or bifid uvula and widespread aortic and arterial aneurysm and tortuosity [1]. This autosomal dominant connective tissue disorder has overlapping clinical features with Marfan syndrome (MFS), including aortic root aneurysm, skeletal features (overgrowth, joint hypermobility), and skin findings. But Loeys-Dietz syndrome (LDS) also has many features that distinguish it from MFS, most prominently clubfoot, craniosynostosis, and cervical spine instability. Overall, the cardiovascular manifestations in LDS are more severe than in MFS, with aortic dissections and ruptures occurring at a younger age and at a smaller diameter. Aortic aneurysms tend to extend beyond the aortic root to involve the aortic branches and cerebral aneurysms. Arterial tortuosity is most commonly present in the neck and head vessels. Although it is reported occasionally, ectopia lentis, which is present in 60% of MFS patients, has not been convincingly linked to the LDS phenotype.

Initially, two clinical subtypes were described: patients diagnosed with LDS type I had characteristic craniofacial features (hypertelorism, cleft palate/bifid uvula, craniosynostosis), whereas no or few craniofacial anomalies were observed in LDS type II patients. The latter were instead characterized by skin abnormalities involving easy bruising and velvety and translucent skin, as can be seen in patients with Ehlers-Danlos syndrome (EDS) [2].

More recently, and taking into account the broad phenotypical spectrum and widened genetic basis that has emerged since the initial description, a new subtype classification of LDS based on the underlying gene has been proposed. LDS types I–V are respectively caused by mutations in *TGFBR1*, *TGFBR2*, *SMAD3*, *TGFB2*, and *TGFB3* [3].

2 CLINICAL FEATURES

2.1 Craniofacial Manifestations

The two most typical craniofacial features are ocular hypertelorism and the presence of a cleft palate, or its mildest presentation, a bifid uvula. The uvula

Aneurysms-Osteoarthritis Syndrome. http://dx.doi.org/10.1016/B978-0-12-802708-0.00007-7
Copyright © 2017 Elsevier Inc. All rights reserved.

can also have an unusually broad appearance with or without a midline raphe. Craniosynostosis is another common presenting feature in the more severely affected patients, and it can involve all sutures—most commonly the sagittal suture (resulting in dolichocephaly), but also the coronal suture (resulting in brachycephaly) and the metopic suture (resulting in trigonocephaly). Other recurrent craniofacial characteristics are malar flattening and retrognathia. Besides the hypertelorism, ocular manifestations include strabismus, blue sclerae, and myopia, but the latter is less frequent and less severe than in patients with MFS. Retinal detachment has rarely been reported. In our experience, ectopia lentis is not observed, although in the literature, minimal lens(sub)luxation has been reported. Less common associated findings requiring further exploration include submandibular branchial cysts and defective tooth enamel.

2.2   Skeletal Manifestations

Marfanoid skeletal features based on bone overgrowth can be observed, although the actual overgrowth is usually milder in LDS patients compared to that of MFS patients. The most striking LDS skeletal findings include pectus deformities, both pectus excavatum and pectus carinatum; scoliosis; joint laxity and contractures; talipes equinovarus; and cervical spine malformation and/or instability. Arachnodactyly is present in about one-third of patients, but true dolichostenomelia (leading to an increase in the arm-span-to-height ratio and a decrease in the upper-to-lower-segment ratio) is less common in LDS patients than in MFS patients. Joint hypermobility is very common and includes congenital hip dislocation and recurrent joint subluxations. Paradoxically, some individuals can show reduced joint mobility, especially of the hands (camptodactyly) and feet (clubfeet). Other recurrent skeletal findings include spondylolisthesis, acetabular protrusion, and pes planum. Preliminary evidence suggests that individuals with LDS have an increased incidence of osteoporosis, with increased fracture incidence and delayed bone healing.

2.3   Cardiovascular Manifestations

In the cardiovascular system, the most common and prominent finding is the dilatation of the aortic root at the sinuses of Valsalva, which, if undetected, can lead to aortic dissection and rupture. Aortic dissections have been described in patients as young as 6 months of age. Importantly, dissections have occurred at smaller diameters than those generally accepted as at risk in MFS patients. In addition to aortic root aneurysms, aneurysms throughout the arterial tree have been described, most prominently in the side branches of the aorta and the cerebral circulation. Finally, another striking finding is the presence of arterial tortuosity, which is usually most prominent in the head and neck vessels. Vertebral and carotid artery dissection and cerebral bleeding have been described. In addition to vascular complications, mitral valve prolapse insufficiency, patent ductus arteriosus, bicuspid aortic valve, and atrial/ventricular septal defects are also more common in patients with LDS.

2.4   Cutaneous Manifestations

In persons without striking craniofacial features, important cutaneous findings can provide the clue toward diagnosis. These skin findings show significant overlap with those observed in patients with EDS and include velvety, thin, translucent skin; easy bruising (other than the lower legs); poor wound healing; and dystrophic scars.

2.5   Other Recurrent Findings

Comparable to the vascular type of EDS, life-threatening complications, such as spontaneous bowel rupture and peripartal uterine rupture, have been reported.

Common neuroradiological findings include dural ectasia and Arnold-Chiari type I malformation, although the precise incidence of those two findings is unknown.

Other recurrent findings that need further documentation include muscle hypoplasia, dental problems with enamel dysplasia, allergic/autoimmune disease with seasonal allergies, asthma/sinusitis, eczema, and important gastrointestinal problems, such as food allergies, eosinophilic esophagitis, and inflammatory bowel disease.

3   DIAGNOSTIC CRITERIA

Although no formal diagnostic criteria have been developed, genetic testing should be considered in the following scenarios:

- Patients with the typical clinical triad of hypertelorism, cleft palate/bifid uvula, and arterial tortuosity/aneurysm
- Early-onset aortic aneurysm with variable combination of other features, including arachnodactyly, camptodactyly, clubfeet, craniosynostosis (all types), blue sclerae, thin skin with atrophic scars, easy bruising, joint hypermobility, bicuspid aortic valve, patent ductus arteriosus, and atrial and ventricular septum defects
- Patients with an MFS-like phenotype, especially those without ectopia lentis, but with aortic and skeletal features not fulfilling the MFS diagnostic criteria
- Families with autosomal dominant thoracic aortic aneurysms, especially those families with early-onset aortic/arterial dissection and aortic disease beyond the aortic root (including cerebral arteries)
- Patients with clinical features reminiscent of vascular EDS (thin skin with atrophic scars, easy bruising, joint hypermobility) and normal type III collagen biochemistry and/or normal *COL3A1* genetic testing
- Sporadic young probands with aortic root dilatation/dissection

4 EXPANDING GENETIC BASIS

Initially, *TGFBR1* and *TGFBR2* were identified as the genes underlying LDS, but more recently *SMAD3*, *TGFB2*, and *TGFB3* have also been identified as genes that can cause LDS (Table 4b.1). Overlapping features are summarized in Table 4b.2.

TABLE 4B.1 Genetic Basis of Loeys-Dietz Syndrome

Gene	Chromosome	Nomenclature	Other names reported
TGFBR1	9q22	LDS, type 1	TAAD2, AAT5, Furlong syndrome
TGFBR2	3p24	LDS, type 2	AAT3, Marfan-like, MFS type 2
SMAD3	15	LDS, type 3	Aneurysms-Osteoarthritis syndrome
TGFB2	1q41	LDS, type 4	Marfan-like syndrome
TGFB3	14q24	LDS, type 5	Rienhoff syndrome

LDS, Loeys-Dietz syndrome; MFS, Marfan syndrome; TGF, transforming growth factor; TAAD, thoracic aortic aneurysms and dissections.

TABLE 4B.2 Overlapping Clinical Features Between the Various Mutations within the Loeys-Dietz Spectrum

	TGFBR1/2	TGFB2	SMAD3	TGFB3
Hypertelorism	+	+	+	+
Bifid uvula/cleft palate	+	+	+	+
Exotropia	+	+	+	+
Craniosynostosis	+	−	−	−
Cervical spine stability	+	−	+	+
Retrognathia surgery	+	+	+	+
Scoliosis/spondylolisthesis	+	+	+	+
Clubfoot	+	+	+	+
Arthrosis	+	−	+	+
Dural ectasia	+	+	+	?
Pneumothorax	+	+	+	−
Hernia	+	+	+	+
Dissection at young age	+	+	+	−
Disease beyond root	+	+	+	+
Cerebral hemorrhage	+	+	+	+
Arterial tortuosity	+	+	+	−
Autoimmune findings	+	+	+	+

Originally, *TGFBR2* and *TGFBR1* mutations were identified in 10 families. Most of these were missense mutations affecting the intracellular serine-threonine kinase domain of the receptors and were shown to lead to loss of function of the receptor activity. However, on the tissue level, a paradoxical upregulation of the TGF-β signaling pathway is found. Subsequently, loss-of-function mutations in *SMAD3*, the first downstream affector, were shown to cause Aneurysms-Osteoarthritis syndrome (AOS; see Chapter 1) [4]. Since the original identification of AOS, many families without osteoarthritis but with LDS-like features, such as hypertelorism, bifid uvula, and craniosynostosis, have been described, putting this entity clearly within the LDS spectrum. Moreover, *SMAD3*-mutation-positive patients seem to have the same severe cardiovascular outcome.

TGFB2 was the first TGF-β ligand involved in aortic disease. Heterozygous mutations, including missense and nonsense mutations and in-frame and frameshift deletions, were originally discovered in a total of 10 families and in two sporadic thoracic aortic aneurysm cases [5]. As the nonsense mutations were predicted to lead to nonsense-mediated decay, loss of function was again hypothesized as the pathogenic mechanism. *TGFB2*-mutation-positive patients also present with a variable clinical phenotype overlapping with LDS. These features include hypertelorism, arterial tortuosity, bicuspid aortic valve, bifid uvula, clubfeet, and easy bruising. The cardiovascular findings in *TGFB2*-mutant patients initially seem to be less severe compared to those of other LDS types, but the full spectrum is still emerging; for example, intracranial aneurysms and subarachnoid hemorrhages have also recently been described in young adults with a *TGFB2* mutation. Despite the predicted loss-of-function effect of these mutations on the TGF-β signaling pathway, an increased canonical and noncanonical signaling pathway in the aortic wall was observed with a shift in ligand use toward TGFB1.

Most recently, another TGF-β ligand, *TGFB3*, was also shown to play a role in syndromic aortic aneurysms and dissections [6]. Typical cardiovascular features present in *TGFB3*-mutant patients include aneurysms and dissections, both occurring in the descending aorta and abdominal aorta, and mitral valve abnormalities, ranging from mild mitral valve prolapse to severe regurgitation with chordae rupture. Typical LDS features encompass hypertelorism, bifid uvula, cervical spine instability, and clubfoot. The mutational spectrum consists of three truncating mutations, an in-frame splice-site mutation, and four missense mutations in a cohort of 43 patients from 11 families. These mutations are primarily located in the latency associated peptide domain and the region of the active cytokine itself. In line with the findings regarding *TGFB2*, it was hypothesized that *TGFB3* mutations would lead to loss of function of TGFB3 but cause a paradoxical increase in TGF-β signaling. Indeed, this hypothesis was confirmed by the demonstration of increased canonical and noncanonical TGF-β signaling and, again, a shift of ligand usage in the aortic wall toward TGFB1.

5 PATHOPHYSIOLOGY

The precise pathophysiological understanding of LDS is complicated by the observation of the so-called TGF-β paradox. Although the mutations in the different genes of the TGF-β signaling pathway (*TGFBR1/2, SMAD3, TGFB2/3*) cause loss of function, it is clear that at the level of the aortic tissue, increased TGF-β signaling is present in the different forms of LDS. The precise cellular mechanisms underlying this uprise remain elusive. For *TGFBR1/2* mutations in LDS, it is becoming increasingly evident that, although the mutations in the serine/threonine kinase domains of these proteins lead to a loss of function of the kinase activity, several lines of evidence confirm that a complete haploinsufficiency of *TGFBR1* or *TGBFR2* does not seem to cause an aortic aneurysm phenotype. First, the currently described nonsense mutations in LDS are all predicted to escape nonsense-mediated decay, and true nonsense mutations in *TGFBR1* cause a skin cancer phenotype, multiple self-healing squamous epithelioma [7]. Second, mouse models with knock-in tgfbr1 or 2 mutations develop aortic aneurysm, whereas haploinsufficiency for tgfbr1 or 2 does not recapitulate the aortic phenotype in mice [8]. As such, expression of mutant tgfbr1 or 2 protein seems to be a prerequisite to cause an aortic aneurysm. However, nonsense mutations and complete deletions of *SMAD3* and *TGFB2/3* suggest that haploinsufficiency or complete loss of function of these genes would be the causal mechanism in those LDS patients. But again, when studying the aortic wall tissue of patients with mutations in *SMAD3* or *TGFB2/3*, a clear upregulation of TGF-β signaling was observed [4–6]. One possible explanation is that upregulation of the noncanonical (ERK/p38/JUN) part of the TGF-β pathway aims to compensate for the loss of the canonical pathway. Alternatively, shifts between isoforms of the TGF-β cytokine family are suggested to play a role. Preliminary evidence suggests that TGFB2- or TGFB3-deficient patients and mice have an increased expression of TGFB1, causing a shift from TGFB2- or -3 to TGFB1-driven signaling. Finally, an important role has been assigned to the difference in embryological origin of the various cell types in the aortic root (eg, second heart field versus neural crest cells). A relative perturbation of TGF-β signaling is predicted to have a disproportionate effect on the more vulnerable lineage (second heart field), resulting in increased ligand expression and excessive TGF-β signaling by adjacent cells of a different lineage (cardiac neural crest), with relative preservation of signaling potential [9].

REFERENCES

[1] Loeys BL, Chen J, Neptune ER, et al. A syndrome of altered cardiovascular, craniofacial, neurocognitive and skeletal development caused by mutations in TGFBR1 or TGFBR2. Nat Genet 2005;37(3):275–81.

[2] Loeys BL, Schwarze U, Holm T, et al. Aneurysm syndromes caused by mutations in the TGF-beta receptor. New Engl J Med 2006;355(8):788–98.

[3] Maccarrick G, Black JH III, Bowdin S, et al. Loeys-Dietz syndrome: a primer for diagnosis and management. Genet Med 2014;16(8):576–87.

[4] van de Laar IM, Oldenburg RA, Pals G, et al. Mutations in SMAD3 cause a syndromic form of aortic aneurysms and dissections with early-onset osteoarthritis. Nat Genet 2011;43(2):121–6.

[5] Lindsay ME, Schepers D, Bolar NA, et al. Loss-of-function mutations in TGFB2 cause a syndromic presentation of thoracic aortic aneurysm. Nat Genet 2012;44(8):922–7.

[6] Bertoli-Avella AM, Gillis E, Morisaki H, et al. Mutations in a TGF-beta ligand, TGFB3, cause syndromic aortic aneurysms and dissections. J Am Coll Cardiol 2015;65(13):1324–36.

[7] Goudie DR, D'Alessandro M, Merriman B, et al. Multiple self-healing squamous epithelioma is caused by a disease-specific spectrum of mutations in TGFBR1. Nat Genet 2011;43(4): 365–9.

[8] Gallo EM, Loch DC, Habashi JP, et al. Angiotensin II-dependent TGF-beta signaling contributes to Loeys-Dietz syndrome vascular pathogenesis. J Clin Invest 2014;124(1):448–60.

[9] Lindsay ME, Dietz HC. Lessons on the pathogenesis of aneurysm from heritable conditions. Nature 2011;473(7347):308–16.

Chapter 4c

Ehlers-Danlos Syndrome

B.L. Loeys, MD, PhD

1 INTRODUCTION

Ehlers-Danlos syndrome comprises a clinically and genetically diverse group of heritable connective tissue disorders that are predominantly characterized by congenital fragility of the connective tissues. The three main organ systems affected by this connective tissue disorder are the skin, joints, and vasculature. Although historically the different forms of Ehlers-Danlos syndrome have been numbered, the current Villefranche nosology recognizes six genetic subtypes based on clinical characteristics, inheritance pattern, and biochemical and molecular findings [1,2]. The overall incidence is estimated to be approximately 1 in 5000 births. The three most common types of Ehlers-Danlos syndrome are the classic, hypermobile, and vascular types, while other types are rather rare. This chapter briefly reviews these different subtypes and also summarizes the knowledge on some recently discovered forms of Ehlers-Danlos like phenotypes. Autosomal dominant, autosomal recessive, and X-linked forms of Ehlers-Danlos syndrome are described (Table 4c.1).

2 CLINICAL PRESENTATION

Ehlers-Danlos syndrome is a heterogeneous group of disorders with variable effects on different organ systems. In the skin, the most common findings are hyperelasticity, atrophic scarring, and thin skin that bruises easily. The joints are typically characterized by hypermobility complicated with recurrent dislocations and chronic pain. But paradoxically in other patients, contractures, clubfeet, and adducted thumbs are also found. Internal organs, most typically the vasculature and the intestines, but also other organs, such as the spleen or the gravid uterus, are prone to spontaneous ruptures. Depending on the subtype, other findings might be diagnostic clues.

3 SUBTYPES OF EHLERS-DANLOS SYNDROME

3.1 Classic Type of Ehlers-Danlos Syndrome (*COL5A1/COL5A2*)

The pathognomic signs in classic Ehlers-Danlos syndrome (formerly known as type I/II) are skin hyperextensibility with the atrophic scarring associated

Aneurysms-Osteoarthritis Syndrome. http://dx.doi.org/10.1016/B978-0-12-802708-0.00008-9
Copyright © 2017 Elsevier Inc. All rights reserved.

TABLE 4C.1 Overview of Ehlers-Danlos subtypes

Type	Gene	Inheritance	Key features
Classic	COL5A1/A2 (COL1A1)	AD	Atrophic scars, skin hyperextensibility, joint hypermobility
TNX-deficient	TNXB	AR	Easy bruising, joint hypermobility, no atrophic scars, skin hyperextensibility
Hypermobile	? Locus chr 8p22-8p21.1	AD	Joint hypermobility, mild skin findings
Vascular	COL3A1	AD	Vascular and intestinal rupture, small joint hypermobility, thin/translucent skin
Periventricular nodular heterotopia	FLNA	XL	Periventricular nodular heterotopia, joint hypermobility, mitral valve disease, aortic aneurysm
Kyphoscoliotic	PLOD1	AR	Kyphoscoliosis, joint hypermobility
Arthrochalasis	COL1A1/A2 exon 6	AD	Congenital hip dislocation, pronounced joint hypermobility
Ehlers-Danlos syndrome/ Osteogenesis Imperfecta	COL1A1/A2 N-terminal	AD	Joint hypermobility, osteopenia
Dermatosparaxis	ADAMTS2	AR	Severe skin fragility with bruising
Cardiac-valvular	COL1A2 null-alleles	AR	Multiple cardiac valve insufficiency, joint hypermobility, skin hyperextensibility
Vascular-like	COL1A1 arg-to-cys	AD	Joint hypermobility with rupture of middle-size arteries, osteoporosis
Progeroid	B4GALT7	AR	Joint hypermobility, loose skin, aged appearance
B3GALT6-deficient	B3GALT6	AR	Spondyloepimetaphyseal dysplasia, bone fragility, muscle hypotonia, kyphoscoliosis, progressive contractures
Musculocontractural or Adducted thumb/clubfoot	CHST14 DSE	AR AR	Congenital joint contractures with joint hypermobility, skin hyperextensibility, kyphoscoliosis
FKBP14-deficient	FKBP14	AR	Sensineural hearing loss, kyphoscoliosis, myopathy
Spondylocheirodysplastic	SLC39A13	AR	Spondyloepiphyseal dysplasia, skin hyperextensibility, wrinkled palms, tapered fingers with contractures
Periodontitis	Locus 12p13	AD	Periodontitis

AD, autosomal dominant; AR, autosomal recessive, XL, X-linked.

with joint hypermobility. The skin usually has a smooth velvety texture as well. Electron microscopy of skin biopsy typically shows a cauliflower-pattern accumulation of collagen molecules and variable diameters of collagen fibrils. Other recurrent findings in patients with classic Ehlers-Danlos syndrome include molluscoid pseudo-tumors, subcutaneous spheroids, muscle hypotonia with delayed gross motor development, easy bruising, and variable manifestations of tissue extensibility and fragility. Aortic root dilatation has also been reported in these patients [3].

This autosomal dominant type of Ehlers-Danlos syndrome is caused by heterozygous mutations in the genes encoding for type V collagen α-chain 1 and 2 (*COL5A1, COL5A2*). These mutations lead to the decreased production or secretion of type V collagen, a regulator of collagen fibril diameter, into the extracellular matrix, establishing haploinsufficiency as the key factor in the pathogenesis of classic Ehlers-Danlos syndrome. About 50% of the patients with classic Ehlers-Danlos syndrome do not show mutations in either of the *COL5* genes [4]. Rarely, some patients with type I collagen (*COL1A1*) mutations have been described [5].

3.2   TNX-Deficient Ehlers-Danlos Syndrome (*TNXB*)

Another rare, recessive type of Ehlers-Danlos syndrome is caused by tenascin X (*TNX*) mutations [6]. Patients with TNX deficiency present with skin hyperextensibility; joint hypermobility, mostly affecting the small joints of the hands; and an extensive history of easy bruising. No major atrophic scarring has been observed in TNX-deficient patients. Some discussion is ongoing as to whether heterozygous carriers of *TNX* mutations also present some degree of joint hypermobility.

3.3   Hypermobile Type of Ehlers-Danlos Syndrome

The major diagnostic criteria of the hypermobile type of Ehlers-Danlos syndrome (formerly known as type III) are generalized joint hypermobility and skin involvement characterized by hyperextensibility and/or smooth velvety skin. Minor diagnostic criteria include recurring joint dislocations, chronic joint/limb pain, and a positive family history. Joint hypermobility is typically weighed through the Beighton score, which evaluates joint hypermobility of the thumb, fingers, elbows, and knees and the patient's ability to place his or her hands flat on the ground.

The genetic basis of joint hypermobility remains largely unknown. Reports of genetic studies in patients/families with Ehlers-Danlos syndrome hypermobility type are scarce. Recently, linkage to locus on chromosome 8p22-8p21.1 has been reported (ASHG 2015, abstract 2831). An occasional missense mutation in the *COL3A1* gene [7] has been reported, but this most likely represents a mild form of vascular Ehlers-Danlos syndrome. Other major collagen genes (types I, II, III, V, VI) have been excluded by linkage studies [8].

A proportion of individuals with the hypermobility type of Ehlers-Danlos syndrome have had aortic root enlargement, but progression of the dilatation associated with these diseases or a predisposition for aortic dissection has not been established [9]. Arguing against the progressive nature of these aortic dilatations is the absence of a history of sudden death in individuals with these disorders.

3.4 Vascular Type of Ehlers-Danlos Syndrome (*COL3A1*)

Typical clinical manifestations of vascular Ehlers-Danlos syndrome (formerly known as type IV) include thin, translucent skin; characteristic facial appearance; vascular fragility demonstrated by extensive bruising and easy bleeding; and spontaneous arterial/intestinal/uterine ruptures [10–12].

Vascular rupture or dissection and gastrointestinal perforation or organ rupture are the presenting signs in 70% of adults. Arterial fragility mostly affects the medium-size arteries. Arterial rupture may be preceded by aneurysm, arteriovenous fistulae, or dissection, or it may occur spontaneously. Neonates may present with clubfoot and/or congenital dislocation of the hips. In childhood, inguinal hernia, pneumothorax, and recurrent joint dislocation or subluxation are common.

Vascular Ehlers-Danlos syndrome is caused by mutations in *COL3A1* (type III collagen α-chain 1) [13]. These mutations consist mostly of missense mutations that lead to the substitution of essential glycine residues within the triple helical domain of the type III collagen chain. Substitution of glycine for valine and splice-donor-site mutations are associated with more severe phenotypes, whereas mutations leading to haploinsufficiency cause milder phenotypes with later onset.

Interestingly, in a cohort of 40 patients displaying a vascular Ehlers-Danlos syndrome-like phenotype but normal collagen III biochemistry, 30% carried *TGFBR1/2* mutations [14], suggesting on the one hand that vascular Ehlers-Danlos syndrome closely resembles Loeys-Dietz syndrome, but on the other hand that *TGFBR* mutations may cause a broad spectrum of diseases associated with aortic aneurysms.

3.5 Ehlers-Danlos Syndrome with Periventricular Nodular Heterotopia (*FLNA*)

Mutations causing loss of function of filamin A (*FLNA*) typically lead to X-linked periventricular nodular heterotopia, with seizures constituting the predominant clinical manifestation of this disorder in female heterozygotes and lethality in males. However, more recently it has been suggested that some patients also present with aortic/arterial dilatation (mainly the aorta), joint hypermobility, and variable skin findings within the spectrum of Ehlers-Danlos. Subsequently, this subgroup has been termed Ehlers-Danlos syndrome—periventricular heterotopia variant [15].

3.6 Kyphoscoliotic Form of Ehlers-Danlos Syndrome (*PLOD1*)

The kyphoscoliotic form of Ehlers-Danlos syndrome (formerly known as type Ehlers-Danlos syndrome VIA) is predominantly characterized by congenital kyphoscoliosis, generalized joint laxity, muscle hypotonia at birth, and, in some individuals, ocular problems (eg, scleral fragility or rupture of the ocular globe, microcornea). Their life spans may be normal, but affected individuals are at risk for rupture of medium-size arteries and respiratory compromise if kyphoscoliosis is severe. Aortic dilation and rupture can also be seen. Other findings include tissue fragility, atrophic scars, easy bruising, marfanoid habitus, and osteopenia.

The kyphoscoliotic form of Ehlers-Danlos syndrome is an autosomal recessive condition caused by deficient activity of the enzyme procollagen-lysine, 2-oxoglutarate 5-dioxygenase 1 (lysyl hydroxylase 1), which is encoded by the *PLOD1* gene [16]. The diagnosis of the kyphoscoliotic form of Ehlers-Danlos syndrome relies on the demonstration of an increased ratio of deoxypyridinoline to pyridinoline crosslinks in urine measured by High Performance Liquid Chromatography, a highly sensitive and specific test. Assay of lysyl hydroxylase enzyme activity in skin fibroblasts is also available.

3.7 Arthrochalasia Type of Ehlers-Danlos Syndrome (*COL1A1, COL1A2*)

Formerly known as Ehlers-Danlos syndrome type VII and now called the arthrochalasia type of Ehlers-Danlos syndrome, this presents with severe generalized joint hypermobility with recurrent subluxations and often congenital bilateral hip dislocation. Associated features are skin hyperextensibility, tissue fragility with atrophic scars, easy bruising, muscle hypotonia, and kyphoscoliosis. Pregnancy is often complicated by polyhydramnios, and bilateral clubfoot is frequently noted on prenatal ultrasounds.

The molecular cause is heterozygous mutations in the exons 6 of *COL1A1* and *COL1A2*, both coding for the procollagen N-proteinase cleavage site [17]. These mutations lead to the loss of the cleavage site for the type I collagen amino-proteinase [18].

3.8 Ehlers-Danlos/Osteogenesis Imperfecta Overlap Syndrome (*COL1A1/COL1A2*)

The Ehlers-Danlos syndrome/osteogenesis imperfecta overlap syndrome results from a mutation in the N-terminal part (up to exon 14) of the type I collagen helical domain (both *COL1A1* and *COL1A2*). The presenting features are usually within the Ehlers-Danlos clinical spectrum, but patients also exhibit mild signs of osteogenesis imperfecta.

Ehlers-Danlos syndrome features include generalized joint hypermobility and dislocations, skin hyperextensibility and/or translucency, easy bruising, and mild abnormal scarring, whereas osteogenesis imperfecta findings include blue sclerae, osteopenia with infrequent fractures, and short stature. Typically, none of these patients are clinically diagnosed with osteogenesis imperfecta, suggesting that the features resulting from bone fragility are less obvious than the features resulting from the soft connective tissue weakness that is characteristic for Ehlers-Danlos syndrome. This autosomal dominant form of Ehlers-Danlos syndrome can also be diagnosed by abnormal migration of the type collagen molecules, which is caused by interference of the mutations with the removal of the N-terminal propeptide [19].

3.9 Dermatosparaxis Type of Ehlers-Danlos Syndrome (*ADAMTS2*)

This autosomal recessive form of Ehlers-Danlos syndrome is characterized by extremely fragile, bruisable, and redundant skin. The joint hypermobility is far less severe than in the arthrochalasia type. The disease is caused by decreased activity of the N-terminal type I collagen propeptides encoded by *ADAMTS2* [20].

3.10 Cardiac-Valvular Type of Ehlers-Danlos Syndrome (*COL1A2*)

The cardiac-valvular type of Ehlers-Danlos syndrome, caused by haploinsufficient mutations in *COL1A2*, is an autosomal recessive form of the syndrome that includes joint hypermobility, skin hyperextensibility, osteopenia and muscular hypotonia in childhood, and severe cardiac valvular defects later in life [21]. This disease is caused by the complete deficiency of the type 1 collagen alpha 2 chain, resulting in the exclusive production of collagen 1 alpha 1 homotrimers.

3.11 Vascular-Like Type of Ehlers-Danlos Syndrome (*COL1A1*)

A very specific subset of arginine-to-cysteine mutations in *COL1A1* (p.Arg312Cys, p.Arg574Cys, and p.Arg1093Cys) has been identified in individuals who typically present with aneurysms of the abdominal aorta and iliac arteries reminiscent of patients with vascular Ehlers-Danlos syndrome. Distinct abnormalities on collagen electrophoresis are observed [22].

Interestingly, other heterozygous arginine-to-cysteine mutations in *COL1A1* mutation (p.Arg888Cys, p.Arg1066Cys) have been identified in patients with the Ehlers-Danlos syndrome/osteogenesis imperfecta overlap condition characterized by osteopenia, skin hyperextensibility without infrarenal aortic, and arterial aneurysms [23].

3.12 Progeroid Type of Ehlers-Danlos Syndrome (*B4GALT7*)

This rare form of Ehlers-Danlos syndrome is characterized by a progeroid appearance in addition to typical Ehlers-Danlos syndrome findings and is caused by mutations in *B4GALT7*, which encodes galactosyltransferase I. Galactosyltransferase I, or β4GalT7, is the enzyme that adds a galactose residue to the O-linked xylose on the proteoglycan core protein [24]. This form of Ehlers-Danlos syndrome is autosomal recessive.

3.13   B3GALT6-Deficient Type Ehlers-Danlos Syndrome (*B3GALT6*)

Interestingly, recessive mutations in the *B3GALT6* gene, encoding galactosyl-transferase II (or β3GalT6), result in an Ehlers-Danlos syndrome-like disorder that presents with spondyloepimetaphyseal dysplasia, bone fragility, muscle hypotonia, kyphoscoliosis, and progressive contractures [25].

3.14   Musculocontractural Type of Ehlers-Danlos Syndrome/ Adducted Thumb-Clubfoot Syndrome (*CHST14, DSE*)

Recessive loss-of-function mutations in *CHST14*, encoding dermatin-4-sulfo-transferase 1 (D4ST1), result in musculocontractural Ehlers-Danlos syndrome (previously known as Ehlers-Danlos syndrome kyphoscoliotic type VIB). This dermatin-4-sulfotransferase 1 catalyzes the 4-O-sulfation of N-acetylgalactos-amine (GalNAc) in the glycosoaminoglycan (GAG) side chain of dermatan sulfate. The clinical spectrum of musculocontractural Ehlers-Danlos syndrome includes significant congenital hypotonia, gross motor delay, muscle hypoplasia with progressive joint hypermobility, progressive early onset kyphoscoliosis, joint contractures, clubfeet, characteristic facial features, thin and bruisable skin, atrophic scarring, and variable ocular involvement [26,27]. Congenital bilateral thumb adduction may be a distinctive clinical finding in this Ehlers-Danlos syndrome subtype. The phenotype overlaps with the kyphoscoliotic type (formerly known as VIA), but distinguishing features include craniofacial abnormalities, joint contractures, wrinkled palms, tapering fingers, and gastro-intestinal and genitourinary manifestations.

In one family, autosomal recessive mutations in the dermatan sulphate epimerase (*DSE*) were found in a family with multisystemic musculocontractural, type 2 [28]. The dermatan sulphate epimerase is required for the epimerization of glucuronic acid (GlcA) to iduronic acid (IdoA).

Both enzyme defects (D4ST1 and DSE) change the disaccharide composition of the GAG chains, eventually leading to a relative lack of dermatan sulphate and excessive chondroitin sulphate in such proteoglycans as versican, thrombomodulin, and small leucine-rich proteoglycans decorin and biglycan.

3.15   FKPB14-Deficient Ehlers-Danlos Syndrome (*FKBP14*)

Mutations in *FKBP14* cause a subtype of Ehlers-Danlos syndrome that re-sembles *CHST14/DSE*-related disorders. Patients present with severe con-genital muscle hypotonia, progressive kyphoscoliosis, joint hypermobility, and hyperelastic skin, but, unlike patients with *CHST14*, these patients also have sensorineural hearing impairment [29,30]. Creatine kinase levels range from normal to mildly elevated (range 60–300 IU/L), nerve conduction studies are normal, and electromyography is normal in infancy but shows a myopathic pattern in adolescence and adulthood. *FKBP14* encodes for an endoplasmatic reticulum protein that belongs to the family FK506 binding isomerases with catalytic function through acceleration of cis-trans isomerization of peptidyl-propyl bonds. *FKBP14* loss-of-function fibroblast analysis shows disturbed distribution and impaired assembly of several extracellular matrix components—specifically, collagen types I and III and fibronectin.

3.16 Spondylocheirodysplastic type of Ehlers-Danlos Syndrome (*SLC39A13*)

The spondylocheirodysplastic form of Ehlers-Danlos syndrome, is characterized by skin hyperextensibility, small-joint hypermobility with tapering fingers, contractures, and mild skeletal dysplasia [31]. Biallelic mutations have been identified in the *SLC39A13* gene, encoding the transmembrane zinc transporter ZIP13, which controls intracellular Zn^{2+} availability. It is hypothesized that deficiency of this transporter leads to decreased activity intracellular Zn^{2+}-dependent processes.

3.17 Periodontitis Type of Ehlers-Danlos Syndrome

This form of autosomal dominant Ehlers-Danlos syndrome has been mapped to a locus on chromosome 12p13, but the causative gene has not been identified yet. The main clinical features are periodontal loss, joint hypermobility, and soft skin with anterior tibial plaques [32].

REFERENCES

[1] Beighton P, de Paepe A, Danks D, et al. International nosology of heritable disorders of connective tissue, Berlin, 1986. Am J Med Genet 1988;29(3):581–94.

[2] Beighton P, De Paepe A, Steinmann B, et al. Ehlers-Danlos syndromes: revised nosology, Villefranche, 1997. Ehlers-Danlos National Foundation (USA) and Ehlers-Danlos Support Group (UK). Am J Med Genet 1998;77(1):31–7.

[3] Monroe GR, Harakalova M, van der Crabben SN, et al. Familial Ehlers-Danlos syndrome with lethal arterial events caused by a mutation in COL5A1. Am J Med Genet A 2015;167(6): 1196–203.

[4] Malfait F, Coucke P, Symoens S, et al. The molecular basis of classic Ehlers-Danlos syndrome: a comprehensive study of biochemical and molecular findings in 48 unrelated patients. Hum Mutat 2005;25(1):28–37.

[5] Mayer SA, Rubin BS, Starman BJ, et al. Spontaneous multivessel cervical artery dissection in a patient with a substitution of alanine for glycine (G13A) in the alpha 1 (I) chain of type I collagen. Neurology 1996;47(2):552–6.

[6] Schalkwijk J, Zweers MC, Steijlen PM, et al. A recessive form of the Ehlers-Danlos syndrome caused by tenascin-X deficiency. New Engl J Med 2001;345(16):1167–75.

[7] Narcisi P, Richards AJ, Ferguson SD, et al. A family with Ehlers-Danlos syndrome type III/articular hypermobility syndrome has a glycine 637 to serine substitution in type III collagen. Hum Mol Genet 1994;3(9):1617–20.

[8] Henney AM, Brotherton DH, Child AH, et al. Segregation analysis of collagen genes in two families with joint hypermobility syndrome. Brit J Rheumatol 1992;31(3):169–74.

[9] Wenstrup RJ, Meyer RA, Lyle JS, et al. Prevalence of aortic root dilation in the Ehlers-Danlos syndrome. Genet Med 2002;4(3):112–7.

[10] Royce PM, Steinmann B. Connective tissue and its heritable disorders: molecular, genetic, and medical aspects. New York: Wiley-Liss; 2002.

[11] Pepin M, Schwarze U, Superti-Furga A, et al. Clinical and genetic features of Ehlers-Danlos syndrome type IV, the vascular type. N Engl J Med 2000;342(10):673–80.

[12] Pepin MG, Schwarze U, Rice KM, et al. Survival is affected by mutation type and molecular mechanism in vascular Ehlers-Danlos syndrome (EDS type IV). Genet Med 2014;16(12): 881–8.

[13] Superti-Furga A, Gugler E, Gitzelmann R, et al. Ehlers-Danlos syndrome type IV: a multi-exon deletion in one of the two COL3A1 alleles affecting structure, stability, and processing of type III procollagen. J Biol Chem 1988;263(13):6226–32.

[14] Loeys BL, Schwarze U, Holm T, et al. Aneurysm syndromes caused by mutations in the TGF-beta receptor. New Engl J Med 2006;355(8):788–98.

[15] Sheen VL, Jansen A, Chen MH, et al. Filamin A mutations cause periventricular heterotopia with Ehlers-Danlos syndrome. Neurology 2005;64(2):254–62.

[16] Pinnell SR, Krane SM, Kenzora JE, et al. A heritable disorder of connective tissue. Hydroxy-lysine-deficient collagen disease. New Engl J Med 1972;286(19):1013–20.

[17] Steinmann B, Tuderman L, Peltonen L, et al. Evidence for a structural mutation of procollagen type I in a patient with the Ehlers-Danlos syndrome type VII. J Biol Chem 1980;255(18): 8887–93.

[18] Giunta C, Chambaz C, Pedemonte M, et al. The arthrochalasia type of Ehlers-Danlos syndrome (EDS VIIA and VIIB): the diagnostic value of collagen fibril ultrastructure. Am J Med Genet A 2008;146A(10):1341–6.

[19] Cabral WA, Makareeva E, Colige A, et al. Mutations near amino end of alpha1(I) collagen cause combined osteogenesis imperfecta/Ehlers-Danlos syndrome by interference with N-propeptide processing. J Biol Chem 2005;280(19):19259–69.

[20] Colige A, Sieron AL, Li SW, et al. Human Ehlers-Danlos syndrome type VII C and bovine dermatosparaxis are caused by mutations in the procollagen I N-proteinase gene. Am J Hum Genet 1999;65(2):308–17.

[21] Schwarze U, Hata R, McKusick VA, et al. Rare autosomal recessive cardiac valvular form of Ehlers-Danlos syndrome results from mutations in the COL1A2 gene that activate the nonsense-mediated RNA decay pathway. Am J Hum Genet 2004;74(5): 917–30.

[22] Malfait F, Symoens S, De Backer J, et al. Three arginine to cysteine substitutions in the pro-alpha (I)-collagen chain cause Ehlers-Danlos syndrome with a propensity to arterial rupture in early adulthood. Hum Mutat 2007;28(4):387–95.

[23] Cabral WA, Makareeva E, Letocha AD, et al. Y-position cysteine substitution in type I collagen (alpha1(I) R888C/p.R1066C) is associated with osteogenesis imperfecta/Ehlers-Danlos syndrome phenotype. Hum Mutat 2007;28(4):396–405.

[24] Okajima T, Fukumoto S, Furukawa K, et al. Molecular basis for the progeroid variant of Ehlers-Danlos syndrome. Identification and characterization of two mutations in galactosyl-transferase I gene. J Biol Chem 1999;274(41):28841–4.

[25] Nakajima M, Mizumoto S, Miyake N, et al. Mutations in B3GALT6, which encodes a glycosaminoglycan linker region enzyme, cause a spectrum of skeletal and connective tissue disorders. Am J Hum Genet 2013;92(6):927–34.

[26] Dündar M, Muller T, Zhang Q, et al. Loss of dermatan-4-sulfotransferase 1 function results in adducted thumb-clubfoot syndrome. Am J Hum Genet 2009;85(6):873–82.

[27] Shimizu K, Okamoto N, Miyake N, et al. Delineation of dermatan 4-O-sulfotransferase 1 deficient Ehlers-Danlos syndrome: observation of two additional patients and comprehensive review of 20 reported patients. Am J Med Genet A 2011;155A(8):1949–58.

[28] Müller T, Mizumoto S, Suresh I, et al. Loss of dermatan sulfate epimerase (DSE) function results in musculocontractural Ehlers-Danlos syndrome. Hum Mol Genet 2013;22(18): 3761–72.

[29] Baumann M, Giunta C, Krabichler B, et al. Mutations in FKBP14 cause a variant of Ehlers-Danlos syndrome with progressive kyphoscoliosis, myopathy, and hearing loss. Am J Hum Genet 2012;90(2):201–16.

[30] Murray ML, Yang M, Fauth C, et al. FKBP14-related Ehlers-Danlos syndrome: expansion of the phenotype to include vascular complications. Am J Med Genet A 2014;164A(7): 1750–5.

[31] Giunta C, Elcioglu NH, Albrecht B, et al. Spondylocheiro dysplastic form of the Ehlers-Danlos syndrome—an autosomal-recessive entity caused by mutations in the zinc transporter gene SLC39A13. Am J Hum Genet 2008;82(6):1290–305.

[32] Rahman N, Dunstan M, Teare MD, et al. Ehlers-Danlos syndrome with severe early-onset periodontal disease (EDS-VIII) is a distinct, heterogeneous disorder with one predisposition gene at chromosome 12p13. Am J Hum Genet 2003;73(1):198–204.

Chapter 4d

Bicuspid Aortic Valve

A.L. Duijnhouwer, MD, A.E. van den Bosch, MD, PhD

1   INTRODUCTION

A bicuspid aortic valve (BAV) is the most common congenital heart condition, with a prevalence of 0.5–2% [1,2]. A BAV is a heterogeneous malformation with a high rate of complications related to the aortic valve and ascending aorta, which occurs in more than 35% of patients with a BAV. Therefore, it is considered as a disease with more significant impact as compared to other congenital cardiac malformations. Depending on the definition and the age group, more than 56% of patients with a BAV have dilatation of the ascending aorta, [3] and in 25 years of follow-up, about 25% need aortic surgery. Presently, there is an ongoing debate regarding the underlying cause of the aortic dilatation in patients with a BAV. The interaction between genetically determined intrinsic disease of the aortic wall, molecular pathways, and altered hemodynamics increasing shear and stretch on the aortic wall all influence the heterogeneous presentation of aortopathy in BAV patients.

2   DEFINITION

The aortic valve normally has three semilunar cusps—a left, a right, and a posterior or noncoronary cusp—housed in a proximal dilation of the aorta called the sinuses of Valsalva, or aortic sinuses (Fig. 4d.1). The left and right cusps are named after the coronary ostia located in these aortic sinuses of Valsalva. A commissure is the area between adjacent cusp where they attach to the aortic sinus wall.

The BAV typically consists of two cusps of unequal size; the larger cusp often has a raphe resulting from fusion of the commissures. The morphology of the BAV can be described in different ways. Sievers et al. described most extensively the exact morphology of the BAV based on pathological specimens (Fig. 4d.2) [4]. The classification is based on fusions between the cusps, presenting at the height of the commissures, also called raphe(s). The spatial orientation or location of the raphes and the functional status of the aortic valve are taken into account. Other classifications also use the orientation of the closing line of the BAV to define three morphological types [5,6].

Aneurysms-Osteoarthritis Syndrome. http://dx.doi.org/10.1016/B978-0-12-802708-0.00009-0
Copyright © 2017 Elsevier Inc. All rights reserved.

73

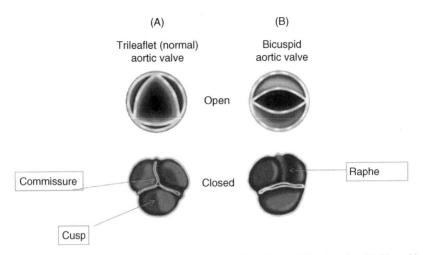

FIGURE 4D.1 Cranial view of the aortic valve. (A) Normal tricuspid aortic valve. (B) Bicuspid aortic valve with a horizontal closing line.

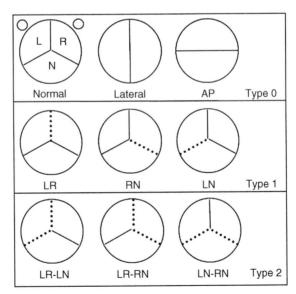

FIGURE 4D.2 Schematic presentation of the bicuspid aortic valve according to the modified Sievers classification. Aortic valves are seen in the short-axis and form a left ventricular view. The *dotted line* represents the raphes (commissural fusion). Type 0 denotes "true" bicuspid aortic valve, which means that there are only two cusps and no raphe. Type 1 and type 2 denote "false" bicuspid aortic valve, which means that there are three cusps and one or two raphes. *R*, right coronary cusp; *L*, left coronary cusp; *N*, noncoronary cusp.

An ascending aortic dilatation is reported in up to 70% of BAV patients. In the majority of BAV patients, this aortic dilatation manifests as an asymmetric ascending aortic dilatation beyond the sinotubular junction with variable arch involvement [3]. Several classifications of aortic root and ascending aorta morphology have been proposed based on the measurements of the sinuses of Valsalva, sinotubular junction, and ascending aorta. Alessandro Della Corte et al. distinguished four types of aortic dilatation based on the aortic ratio (the measured aortic diameter divided by the expected aortic diameter): normal, small, mid-ascending phenotype, and root phenotype [3]. Shafie Fazel et al. distinguished four types of aorta dilatation by clustering the most common dilated aorta parts: (1) solely sinus of Valsalva dilatation (13%); (2) solely ascending aortic dilatation (14%); (3) ascending and transverse aortic arch dilatation (28%); and (4) diffuse dilatation of aorta sinus of Valsalva, ascending aorta, and transverse aortic arch (45%) [7]. The most practical classification was made by Benjamin Schaefer et al. [5] (Fig. 4d.3).

Hector Michelena et al. found that BAV patients have an eight-fold increased risk of aortic dissection, a 26% risk of aneurysm formation, and a 25%

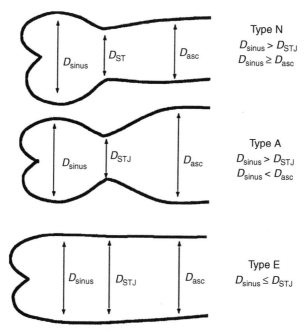

Type N
$D_{sinus} > D_{STJ}$
$D_{sinus} \geq D_{asc}$

Type A
$D_{sinus} > D_{STJ}$
$D_{sinus} < D_{asc}$

Type E
$D_{sinus} \leq D_{STJ}$

FIGURE 4D.3 **Schematic presentation of the spectrum of possible patterns of aortic dimensions observed with bicuspid aortic valves and their classification according to Schaefer.** *D,* diameter; *STJ,* sinotubular junction; *asc,* ascending aorta. *(Reprinted with permission from the article by Schaefer BM, Lewin MB, Stout KK, Gill E, Prueitt A, Byers PH, et al. The bicuspid aortic valve: an integrated phenotypic classification of leaflet morphology and aortic root shape. Heart 2008;94(12):1634–1638 [5].)*

chance for aortic surgery, emphasizing the condition's health burden [8]. In this same study, no dissections occurred in patients with a baseline aortic diameter smaller than 45 mm or with normally functioning aortic valves after 16 years of follow-up, information that was confirmed in other study [9]. The International Registry of Acute Aortic Dissections (IRAD) reported type A dissection (according to Standford classification) in 617 patients of which 4% had a BAV and type B dissection in 384 patients of which 2% had a BAV [10]. In patients under the age of 40 years, the percentage with a BAV increases to 9% ($n = 6$). The Genetically Triggered Thoracic Aortic Aneurysms and Cardiovascular Conditions (GenTAC) Registry Consortium reports a 7% aortic dissection rate in patients with a BAV and a thoracic aorta aneurysm [11].

3 EMBRYOLOGY OF THE AORTIC VALVE AND PROXIMAL AORTA

The embryonic truncus arteriosus is formed from neural crest cells. It starts when neural crest cells migrate over the pharyngeal arches 4 and 6 to the location at which the truncus arteriosus and the conus cordis are formed. These neural crest cells transform into mesenchymal cells that proliferate into truncoconal cushions. The two truncoconal ridges (neural crest cells and cells originating from the second heart field) grow toward each other and fuse first at the truncoconal transition before closing distally (toward the outflow tract) and proximally (toward the ventricles). Closing/zipping of the ridges happens in a spiral motion, in such a way that the pulmonary artery ends up anterior of the aorta.

The aortic semilunar valves are formed out of a cavitation of the truncoconal ridge tissue. This formation happens in a stereotypical way: the normal right and left aortic cusps form at the junction of the ventricular and arterial ends of the conotruncal channel (cells originating from neural crest cells). The noncoronary cusp normally forms from additional conotruncal channel tissue (cells originating from a second heart field). Abnormalities in the area of the conotruncal channel lead to the development of a BAV, often with incomplete separation (or fusion) of the valve tissue.

4 GENES INVOLVED IN THE MALFORMATION OF THE CONOTRUNCUS

The inheritance of a BAV is presumed to be autosomal dominant with incomplete penetrance and is approximately three times more frequent in males than in females. To date, most genes involved in the formation of the conotruncal channel are still unknown, although some mutations have been associated with BAV formation.

An important gene in the formation of the conotruncal channel is the *NOTCH1* gene, which is essential for encoding for the signaling receptor required throughout the development of the aortic valve. Mutations in the *NOTCH1* gene

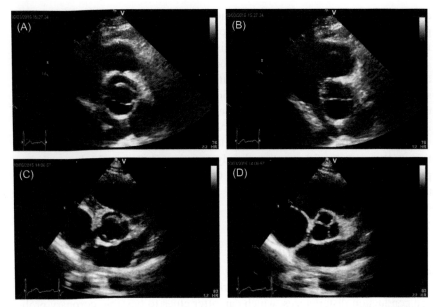

FIGURE 4D.4 **Echocardiographic parasternal short-axis view of two types of bicuspid aortic valve.** (A) systolic phase, valve open with (B) closed valve, of a type 0 bicuspid aortic valve according to the Sievers classification. (C) Systolic phase, open valve with (D) closed valve of a type 1, LR. This demonstrates that the closed view is important for determining the exact Sievers type.

or vertical). In combination with the parasternal long-axis view, the echo density and thickness of the valve cusps, their mobility, and their coaptation can be assessed (Fig. 4d.4). Based on previous description, BAV can be classified [4] (Fig. 4d.4). The advantage of using Sievers's classification is that the morphology of the bicuspid valve has its own typing. In more than 70% of patients, fusion of the right and left coronary cusps is present. Describing the morphology of the BAV is important, because some morphology types have a higher likelihood of complications [4]. In a phenotype that includes the noncoronary cusp, there is a higher likelihood of ascending aorta dilatation.

- Assessment of aortic valve function does not differ from other types of aortic valve disease [31]. Quantification of aortic valve stenosis includes measurements of maximum aortic jet velocity, calculation of mean transaortic pressure gradient, and determination of aortic valve area.
- Assessment of aortic valve regurgitation is performed by measurement of the vena contracta, continues wave pressure half time and presence of backflow in the descending and abdominal aorta, requires some more experience [33] and in this regard indirect signs (eg, left ventricle dilatation, back flow in aortic arch and abdomen) of severe regurgitation are sometimes all there is to distinguish between severe and moderate regurgitation, especially in eccentric regurgitations jets.

lead to signaling abnormalities that may cause an abnormal development of the aortic valve, resulting in a BAV [12]. Mutations in the *NOTCH1* gene have been associated with dominantly inherited BAV in a small number of families.

Another possible etiology is mutations in the transforming growth factor-beta receptor 2 gene (*TGFBR2*), which were found in a patient with a BAV and an aortic aneurysm. However, other studies reported no mutations in these genes. *GATA5* gene variations have been related to bicuspid valve formation, as opposed to *TGFBR2* gene mutation and *NOTCH1* gene mutations; variations in these genes do not cause left ventricle outflow tract malformations.

In humans, endothelial nitric oxide synthase (eNOS) expression is lower in the ascending aorta with a BAV compared to tricuspid aortic valves. During normal cardiogenesis, eNOS expression is restricted to endocardial cells and is shear stress-dependent. Endothelium-derived nitric oxide is known to mediate endothelial cell podokinesis [13]. eNOS plays a role in the formation of valves and vascular remodeling; eNOS deficiency therefore may lead to BAVs. It is probable that all BAV type morphologies are different etiological entities, but this has not been confirmed in humans.

5 THE RELATIONSHIP BETWEEN AORTIC DILATATION AND BICUSPID AORTIC VALVE

In patients with a BAV, the aortic root and ascending aorta are significantly larger than those in patients with a tricuspid aortic valve, even in BAV patients with normal valve function. Controversies exist in the literature regarding the underlying pathogenesis of BAV-associated thoracic aortopathy—specifically, whether it is genetic or hemodynamic in origin. It seems that both contribute to the aortopathy. The evidence supporting a genetic cause includes aortopathy in first-degree relatives of BAV patients and larger aorta diameters in BAV patients, even after correction of possible confounders. Moreover, it is known that both the aortic and the main pulmonary arteries are often dilated in BAV patients, which suggests a common developmental abnormality, as the great arteries originate from a common embryonic trunk and abnormal migration of neural-crest cells.

Yet some evidence supports a hemodynamic cause. For example, even normal functioning BAVs have abnormal transvalvular-flow patterns, resulting in an increased shear wall stress. Different bicuspid morphologies seem to cause different types of aorta dilatations. However, valve morphology is not a predictor of future events [8,14,15]. The histologic changes observed in BAV-associated thoracic aortopathy (cystic medial necrosis) are the end result of abnormal regulatory pathways of the aortic wall [16]. These changes can be seen in many others causes of aortopathy (eg, Marfan syndrome, Turner syndrome). The susceptibility of the aortic wall is probably partially genetic in origin and partially caused by external factors (eg, shear wall stress, high intra-aortic pressure, inflammatory triggers).

6   CLINICAL PRESENTATION OF PATIENTS WITH A BICUSPID AORTIC VALVE

In the pediatric population, most BAVs are detected during regular pediatric checkup and/or as a secondary diagnosis with another malformation. In our unpublished data on 251 BAV patients, the mean age was 5.8 years, the female-to-male ratio was 2:1, and BAVs occurred in conjunction with other cardiac malformations in 59% of the subjects. The most common coexisting malformations were coarctation of the aorta (34%), ventricular septal defect (23%), and persistent ductus arteriosus (15%). A BAV in association with syndromes were seen in 11.6% ($n = 29$), with Turner syndrome, 22Q11 deletion, and Down syndrome being most prevalent. These data are comparable with the published data [17–19]. In this population, only 15% of the patients received a valve intervention. Other possible associated malformations and/or syndromes include hypoplastic left heart syndrome, interruption of the aortic arch, Shone's syndrome, Williams syndrome, and Turner syndrome [20–23].

Coronary artery anomalies (eg, single coronaries, reversal of coronary dominance) are suggested, but no convincing evidence shows that the prevalence in BAV patients is higher than in other patients [24–26].

In adulthood, the mode of BAV presentation has not been specifically investigated until now, but a part of the adult population will flow from the child population. Some patients are diagnosed coincidentally—for example, during investigations of other health concerns—but an increasing number is found in the context of screening first-degree family relatives.

7   EVALUATION OF PATIENTS WITH BICUSPID AORTIC VALVE

7.1   History

Most BAV patients are asymptomatic and have normal exercise tolerance. When symptoms occur, they are mostly related to valve dysfunction and are not different in BAV patients than in patients with other causes of aortic valve stenosis or insufficiency. Aortic root dilatation almost never gives rise to complaints, and when present, it mostly manifests in atypical chest and back pains. Aortic dissection is often the first presentation of aortic dilatation in BAV patients.

Family history is an important part of the evaluation for patients and for their families. Sudden cardiac death and dissection in family history are risk factors for future dissection. A positive family history for BAV, aortic dissection, aortic surgery, and/or aortic valve surgery suggests inheritability. It is probably reasonable to screen first-degree family members in these circumstances.

7.2   Physical Examination

Next to the standard physical examination, a few important aspects are worth highlighting. High blood pressure is probably the most important cause of dilatation and dissection of the aorta. Strict regulation of blood pressure is mandatory to prevent fast aortic dilatation and, thus, aortic surgery or dissection. According to IRAD, 72.1% of all patients with aortic dissection also had hypertension [27].

Obesity expressed as a measure of body mass index $> 25 \text{ kg/m}^2$ is a disease associated with developed countries, with many related conditions—hypertension, immobility, poor physical condition, diabetes, and high cholesterol are only a few. Making patients aware of the risks associated with obesity and giving lifestyle advice is an important intervention.

Physical examination contributes little to the diagnosis of BAV, aortic dilatation, or the severity of the aortic valve dysfunction. But a standard exam is important for getting a general impression of the patient, in case future operations and interventions become necessary.

7.3   Additional Testing

7.3.1   Electrocardiography

Electrocardiography (ECG) is a widely used investigative tool within cardiology and is important in the general work-up of BAV patients to detect rhythm and conduction disturbances, which can offer clues to possible coronary problems and left ventricular hypertrophy. Left ventricular hypertrophy shown on an ECG is a cardiovascular risk factor independent of left ventricular hypertrophy shown on an echocardiogram, as proven in patients > 70 years [28].

7.3.2   Laboratory Testing

Laboratory testing is not routinely advised. In patients with symptomatic aortic valve stenosis or aortic valve regurgitation, the natriuretic peptide is elevated. The additional prognostic value of elevated natriuretic peptide is still unknown for this patient population [29].

7.3.3   Echocardiography

When diagnosing and following up with BAV patients, echocardiography is the most important imaging tool for evaluating the aortic valve [30,31]. Echocardiography provides detailed anatomic data and valve function. Echocardiographic investigation should be extensive to exclude other cardiac malformations; some specific measurements are highlighted here in the context of BAV patients:

- Left ventricular function expressed as a left ventricular ejection fraction (LVEF) should be measured, for it is an important prognostic factor. Three-dimensional echocardiography is the preferred method for LVEF measurement, as this technique has proven to be more accurate than two-dimensional echocardiography [32].
- Aortic valve morphology can be assessed in the parasternal short axis view. The morphology description should entail the presence or absence of a raphe, the raphe's location, and the orientation of the commissure (horizontal

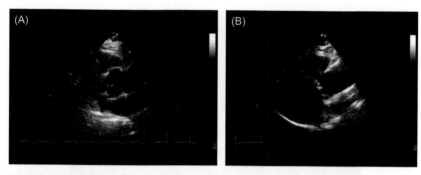

FIGURE 4D.5 Echocardiographic parasternal long-axis view. (A) Type N proximal aortic dilatation according to Schaefer classification. (B) Type E proximal aortic dilatation according to Schaefer classification.

- Visualization of the aortic root in adults is best performed from a parasternal long-axis standard view and, for the ascending aorta, one rib higher, or the right parasternal long-axis view. Measurements should be done from leading edge to leading edge [34]. Important landmarks of the aorta root include aortic annulus, sinuses of Valsalva, sinotubular junction, and proximal ascending aorta.
- Describing the shape of the aorta root may also useful. Schaeffer et al. described one of the practical classifications (Fig. 4d.5) and revealed a possible association between valve morphology and type of aortic root dilatation [5].

Transesophageal echocardiography (TEE) can be helpful, especially in patients with a poor transthoracic echocardiographic (TTE) window. When in doubt about the aortic valve's function during TTE, TEE offers additional value for the assessment of aortic valve morphology and function; for aortic dimensions, the cardiac-MRI and cardiac-CT are more accurate and less invasive.

7.3.4   Cardiac-MRI and Cardiac-CT

Both modalities are suitable for evaluating the thoracic aorta. MRI has the advantage of not requiring radiation, and thus its use is theoretically unlimited. Also, functional imaging can be required using cardiac-MRI (eg, left ventricular volume and flow velocity measurements). Disadvantages of cardiac-MRI include long acquisition time and the difficulty of the technique, which requires highly specialized technicians, making it also expensive. The resolution of an MRI is lower compared to that of a cardiac-CT, but it can be enhanced by using gadolinium-based MRI contrast. An obvious disadvantage of cardiac-CT is its use of radiation. For accurate measurement of the aortic annulus, sinuses of Valsalva, and ascending aorta, the CT-scan should be ECG-triggered, because these anatomical sites move during the cardiac cycle and expand during systole. With respect to thoracic aorta dimensions with MRI (Fig. 4d.6) or CT, it is important that the measurements be taken at reproducible landmarks in diastole [35] and

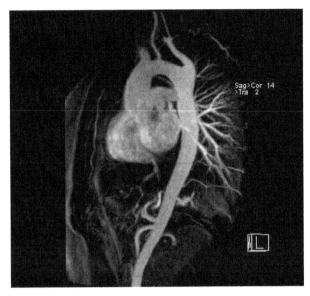

FIGURE 4D.6   MR angiography of the aorta: ascending aorta dilatation and unexpected dilatation of transition aortic arch and descending aorta, below the left subclavian artery. This demonstrates the importance of MR-imaging of the full aorta.

from external border to external border perpendicular to the aortic wall. This requirement differs from echocardiography.

Next to the absolute measurements, it is always important to look at the images produced of the aorta with MRI or CT, because important malformation and/or complications can be detected, albeit it rarely (Fig. 4d.7). Aortic arch anomalies occur in BAV patients, and some evidence suggests that a so-called bovine arch (common origin of the brachiocephalic artery and left common carotid artery) is a predictor of dissection [36]. Additional measurements of the left and right ventricle volumes can be obtained with an MRI. With ECG-triggered CT, these volumes can be measured, but at the cost of much more radiation.

7.3.5   Cardiopulmonary Exercise Testing

Cardiopulmonary exercise testing is preeminently suitable for young patients, because patients in this group develop symptoms late due to their commonly large physical reserve. The reproducibility of cardiopulmonary exercise testing contributes to the early detection of decline in exercise capacity caused by a reduction in stroke volume and detection of the presymptomatic period.

After a full evaluation of the BAV patient, a risk stratification for ascending aorta dissection could be done by evaluating the diameters of the complete ascending aorta and aortic sinuses of Valsalva and their progression over time.

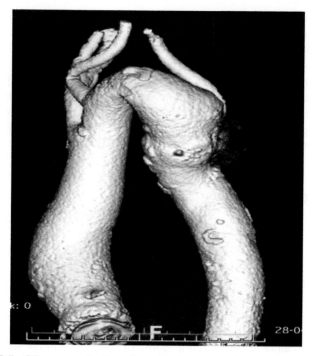

FIGURE 4D.7 CT angiography three-dimensional reconstruction, showing a small suture aneurysm and the cervical course of the aortic arch, with acute angle and dilatated ascending aorta.

Fast growth > 5 mm/year is associated with an increased risk of dissection [37]. Also, patients with a history of coarctation aorta [38,39] and hypertension and a family history of dissection of the aorta have a higher dissection risk.

8 TREATMENT

8.1 Follow-Up

The American College of Cardiology (ACC) 2014 guidelines and the European Society of Cardiology (ESC) 2012 guidelines [40,41] on valvular heart disease recommend TTE in patients with known BAV to evaluate valve morphology, measure the severity of the aortic valve stenosis and regurgitation, and assess the diameter of the aortic sinuses of Valsalva and the ascending aorta. The ACC guidelines also recommend evaluating the shape of the aortic sinuses of Valsalva and the ascending aorta to predict clinical outcome and to determine the timing of an intervention. Determining the BAV morphology is recommended, because fusion involving the noncoronary cusp (Sievers type 1, RN or LN) is associated with a higher incidence of aortic dilatation. This dilatation is more likely to occur in the ascending aorta.

Serial measurements are recommended by guidelines and interval, depending on aorta diameter, aortic growth rate, and family history of dissection. Annual evaluation of the aorta root and ascending aorta is recommended if any aortic diameter is > 45 mm or has a growth rate of > 3 mm/year.

Both guidelines recommend MRI angiography or CT angiography if the aortic sinuses of Valsalva, sinotubular junction, or ascending aorta cannot be assessed accurately or fully by TTE. The ESC guidelines also advise performing MRI angiography or CT angiography if the aortic diameter at any level is > 50 mm to confirm the measurement of the TTE.

8.2   Surgical Intervention

The most recent sets of ACC and ESC guidelines were adjusted with respect to timing of the surgical intervention. Thoracic aortic dilatation is more prevalent in BAV patients, but despite this higher growth rate, there is no evidence that the aorta of a BAV patient has a higher likelihood of complications than that of a patient with a tricuspid aortic valve. Surgical repair or replacement of the aortic sinuses of Valsalva or ascending aorta is indicated in BAV patients if the aortic diameter is > 55 mm, or 27.5 mm/m^2 for patients with a short stature. In patients with an aortic diameter >50 mm, surgical intervention can be considered if a history of coarctation of the aorta, a family history of aorta dissection, or a fast growth rate (> 5 mm/year) is present. Replacement of the ascending aorta is reasonable in BAV patients with who are undergoing aortic valve surgery if the diameter of the ascending aorta is > 45 mm.

Both guidelines give no specific recommendations about the frequency of follow-up in patients with aortic root or ascending aorta < 45 mm. Most patients will have some sort of aortic valve dysfunction, and guidelines for their specific dysfunctions can be applied. In newly diagnosed BAV patients, it seems reasonable to initially evaluate them annually if their aortic valve function is normal or is only mildly to moderately dysfunctional. If after an initial 2–3 years of annual check-ups the aortic valve function has not changed and the aortic diameter has remained stable (with preferably at least 1 MRI angiography or CT angiography to confirm aortic diameters), a follow-up frequency of every 2–4 years is reasonable and safe, especially in patients who comprehend instructions about how to recognize symptoms.

In BAV patients with morphology that involves a raphe with the noncoronary cusp and/or moderate to severe aortic valve regurgitation, hypertension, coarctation of the aorta, and family history of dissection, a second MRI angiography after at least 5 years to check for ascending aorta and proximal arch dilatation should be considered; previously mentioned disease-modifying factors can cause faster dilatation of the ascending aorta and proximal aortic arch.

lead to signaling abnormalities that may cause an abnormal development of the aortic valve, resulting in a BAV [12]. Mutations in the *NOTCH1* gene have been associated with dominantly inherited BAV in a small number of families.

Another possible etiology is mutations in the transforming growth factor-beta receptor 2 gene (*TGFBR2*), which were found in a patient with a BAV and an aortic aneurysm. However, other studies reported no mutations in these genes. *GATA5* gene variations have been related to bicuspid valve formation, as opposed to *TGFBR2* gene mutation and *NOTCH1* gene mutations; variations in these genes do not cause left ventricle outflow tract malformations.

In humans, endothelial nitric oxide synthase (eNOS) expression is lower in the ascending aorta with a BAV compared to tricuspid aortic valves. During normal cardiogenesis, eNOS expression is restricted to endocardial cells and is shear stress-dependent. Endothelium-derived nitric oxide is known to mediate endothelial cell podokinesis [13]. eNOS plays a role in the formation of valves and vascular remodeling; eNOS deficiency therefore may lead to BAVs. It is probable that all BAV type morphologies are different etiological entities, but this has not been confirmed in humans.

5   THE RELATIONSHIP BETWEEN AORTIC DILATATION AND BICUSPID AORTIC VALVE

In patients with a BAV, the aortic root and ascending aorta are significantly larger than those in patients with a tricuspid aortic valve, even in BAV patients with normal valve function. Controversies exist in the literature regarding the underlying pathogenesis of BAV-associated thoracic aortopathy—specifically, whether it is genetic or hemodynamic in origin. It seems that both contribute to the aortopathy. The evidence supporting a genetic cause includes aortopathy in first-degree relatives of BAV patients and larger aorta diameters in BAV patients, even after correction of possible confounders. Moreover, it is known that both the aortic and the main pulmonary arteries are often dilated in BAV patients, which suggests a common developmental abnormality, as the great arteries originate from a common embryonic trunk and abnormal migration of neural-crest cells.

Yet some evidence supports a hemodynamic cause. For example, even normal functioning BAVs have abnormal transvalvular-flow patterns, resulting in an increased shear wall stress. Different bicuspid morphologies seem to cause different types of aorta dilatations. However, valve morphology is not a predictor of future events [8,14,15]. The histologic changes observed in BAV-associated thoracic aortopathy (cystic medial necrosis) are the end result of abnormal regulatory pathways of the aortic wall [16]. These changes can be seen in many others causes of aortopathy (eg, Marfan syndrome, Turner syndrome). The susceptibility of the aortic wall is probably partially genetic in origin and partially caused by external factors (eg, shear wall stress, high intra-aortic pressure, inflammatory triggers).

6 CLINICAL PRESENTATION OF PATIENTS WITH A BICUSPID AORTIC VALVE

In the pediatric population, most BAVs are detected during regular pediatric checkup and/or as a secondary diagnosis with another malformation. In our unpublished data on 251 BAV patients, the mean age was 5.8 years, the female-to-male ratio was 2:1, and BAVs occurred in conjunction with other cardiac malformations in 59% of the subjects. The most common coexisting malformations were coarctation of the aorta (34%), ventricular septal defect (23%), and persistent ductus arteriosus (15%). A BAV in association with syndromes were seen in 11.6% ($n = 29$), with Turner syndrome, 22Q11 deletion, and Down syndrome being most prevalent. These data are comparable with the published data [17–19]. In this population, only 15% of the patients received a valve intervention. Other possible associated malformations and/or syndromes include hypoplastic left heart syndrome, interruption of the aortic arch, Shone's syndrome, Williams syndrome, and Turner syndrome [20–23].

Coronary artery anomalies (eg, single coronaries, reversal of coronary dominance) are suggested, but no convincing evidence shows that the prevalence in BAV patients is higher than in other patients [24–26].

In adulthood, the mode of BAV presentation has not been specifically investigated until now, but a part of the adult population will flow from the child population. Some patients are diagnosed coincidentally—for example, during investigations of other health concerns—but an increasing number is found in the context of screening first-degree family relatives.

7 EVALUATION OF PATIENTS WITH BICUSPID AORTIC VALVE

7.1 History

Most BAV patients are asymptomatic and have normal exercise tolerance. When symptoms occur, they are mostly related to valve dysfunction and are not different in BAV patients than in patients with other causes of aortic valve stenosis or insufficiency. Aortic root dilatation almost never gives rise to complaints, and when present, it mostly manifests in atypical chest and back pains. Aortic dissection is often the first presentation of aortic dilatation in BAV patients.

Family history is an important part of the evaluation for patients and for their families. Sudden cardiac death and dissection in family history are risk factors for future dissection. A positive family history for BAV, aortic dissection, aortic surgery, and/or aortic valve surgery suggests inheritability. It is probably reasonable to screen first-degree family members in these circumstances.

7.2 Physical Examination

Next to the standard physical examination, a few important aspects are worth highlighting. High blood pressure is probably the most important cause of dilatation and dissection of the aorta. Strict regulation of blood pressure is mandatory to prevent fast aortic dilatation and, thus, aortic surgery or dissection. According to IRAD, 72.1% of all patients with aortic dissection also had hypertension [27].

Obesity expressed as a measure of body mass index > 25 kg/m^2 is a disease associated with developed countries, with many related conditions—hypertension, immobility, poor physical condition, diabetes, and high cholesterol are only a few. Making patients aware of the risks associated with obesity and giving lifestyle advice is an important intervention.

Physical examination contributes little to the diagnosis of BAV, aortic dilatation, or the severity of the aortic valve dysfunction. But a standard exam is important for getting a general impression of the patient, in case future operations and interventions become necessary.

7.3 Additional Testing

7.3.1 Electrocardiography

Electrocardiography (ECG) is a widely used investigative tool within cardiology and is important in the general work-up of BAV patients to detect rhythm and conduction disturbances, which can offer clues to possible coronary problems and left ventricular hypertrophy. Left ventricular hypertrophy shown on an ECG is a cardiovascular risk factor independent of left ventricular hypertrophy shown on an echocardiogram, as proven in patients > 70 years [28].

7.3.2 Laboratory Testing

Laboratory testing is not routinely advised. In patients with symptomatic aortic valve stenosis or aortic valve regurgitation, the natriuretic peptide is elevated. The additional prognostic value of elevated natriuretic peptide is still unknown for this patient population [29].

7.3.3 Echocardiography

When diagnosing and following up with BAV patients, echocardiography is the most important imaging tool for evaluating the aortic valve [30,31]. Echocardiography provides detailed anatomic data and valve function. Echocardiographic investigation should be extensive to exclude other cardiac malformations; some specific measurements are highlighted here in the context of BAV patients:

- Left ventricular function expressed as a left ventricular ejection fraction (LVEF) should be measured, for it is an important prognostic factor. Three-dimensional echocardiography is the preferred method for LVEF measurement, as this technique has proven to be more accurate than two-dimensional echocardiography [32].
- Aortic valve morphology can be assessed in the parasternal short axis view. The morphology description should entail the presence or absence of a raphe, the raphe's location, and the orientation of the commissure (horizontal

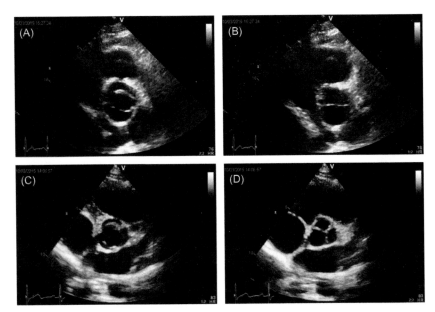

FIGURE 4D.4 **Echocardiographic parasternal short-axis view of two types of bicuspid aortic valve.** (A) systolic phase, valve open with (B) closed valve, of a type 0 bicuspid aortic valve according to the Sievers classification. (C) Systolic phase, open valve with (D) closed valve of a type 1, LR. This demonstrates that the closed view is important for determining the exact Sievers type.

or vertical). In combination with the parasternal long-axis view, the echo density and thickness of the valve cusps, their mobility, and their coaptation can be assessed (Fig. 4d.4). Based on previous description, BAV can be classified [4] (Fig. 4d.4). The advantage of using Sievers's classification is that the morphology of the bicuspid valve has its own typing. In more than 70% of patients, fusion of the right and left coronary cusps is present. Describing the morphology of the BAV is important, because some morphology types have a higher likelihood of complications [4]. In a phenotype that includes the noncoronary cusp, there is a higher likelihood of ascending aorta dilatation.

- Assessment of aortic valve function does not differ from other types of aortic valve disease [31]. Quantification of aortic valve stenosis includes measurements of maximum aortic jet velocity, calculation of mean transaortic pressure gradient, and determination of aortic valve area.
- Assessment of aortic valve regurgitation is performed by measurement of the vena contracta, continues wave pressure half time and presence of backflow in the descending and abdominal aorta, requires some more experience [33] and in this regard indirect signs (eg, left ventricle dilatation, back flow in aortic arch and abdomen) of severe regurgitation are sometimes all there is to distinguish between severe and moderate regurgitation, especially in eccentric regurgitations jets.

Especially with TTE, the distal part of the ascending aorta and proximal arch cannot be fully assessed.

8.3 Medical Therapy

No proven medical therapy is available to halt aortic valve stenosis progression or ascending aorta dilatation. Statins have been investigated but with different outcomes on aortic valve stenosis progression. Adequate treatment of hypertension in BAV patients is important, because more than 70% of patients with aortic dissection have hypertension. Prophylaxis for endocarditis is not indicated.

8.4 Family Screening and Exercise

The current ACC and ESC guidelines advise doing a first-degree family screening for BAV and associated malformations, knowing that 10–20% of BAV patients also have a family member with BAV [42].

Advice regarding sports activities is difficult, because no real evidence exists either way. It seems reasonable to advise against high isotonic contractions, especially in patients with an aortic diameter > 45 mm.

8.5 Prognosis

When BAV patients are treated according to the current guidelines, their prognosis is favorable and similar to that of an age-matched normal population [43].

8.6 Management During Pregnancy

According to the 2011 ESC guidelines on cardiovascular heart disease and pregnancy, [44] BAV patients have an increased risk of complications during pregnancy, depending on the aorta diameter (Table 4d.1).

TABLE 4D.1 Aorta Diameter and the Maternal Pregnancy Risk

Aortic diameter	WHO classification
< 45 mm	2
Between 45–50 mm	3
> 50 mm	4

WHO classification 2 means a small increased risk of maternal mortality or moderate increase in morbidity; WHO classification 3 means a significantly increased risk of maternal mortality or severe morbidity (expert counseling required); WHO classification 4 means an extremely high risk of maternal mortality or severe morbidity, pregnancy contraindicated.

REFERENCES

[1] Basso C, Boschello M, Perrone C, Mecenero A, Cera A, Bicego D, et al. An echocardiographic survey of primary school children for bicuspid aortic valve. Am J Cardiol 2004;93(5):661–3.

[2] Hoffman JI, Kaplan S. The incidence of congenital heart disease. J Am Coll Cardiol 2002;39(12):1890–900.

[3] Della Corte A, Bancone C, Quarto C, Dialetto G, Covino FE, Scardone M, et al. Predictors of ascending aortic dilatation with bicuspid aortic valve: a wide spectrum of disease expression. Eur J Cardio-Thorac 2007;31(3):397–404.

[4] Sievers H-H, Schmidtke C. A classification system for the bicuspid aortic valve from 304 surgical specimens. J Thorac Cardiov Sur 2007;133(5):1226–33.

[5] Schaefer BM, Lewin MB, Stout KK, Gill E, Prueitt a, Byers PH, et al. The bicuspid aortic valve: an integrated phenotypic classification of leaflet morphology and aortic root shape. Heart 2008;94(12):1634–8.

[6] Roberts W. The congenitally bicuspid aortic valve. A study of 85 autopsy cases. Am J Cardiol 1970;26(1):72–83.

[7] Fazel SS, Mallidi HR, Lee RS, Sheehan MP, Liang D, Fleischman D, et al. The aortopathy of bicuspid aortic valve disease has distinctive patterns and usually involves the transverse aortic arch. J Thorac Cardiov Sur 2008;135(4):901–7.

[8] Michelena HI, Khanna AD, Mahoney D, Margaryan E, Topilsky Y, Suri RM, et al. Incidence of aortic complications in patients with bicuspid aortic valves. J Am Coll Cardiol 2011;306(10):1104–12.

[9] Borger MA, Preston M, Ivanov J, Fedak PW, Davierwala P, Armstrong S, David TE. Should the ascending aorta be replaced more frequently in patients with bicuspid aortic valve disease? J Thorac Cardiovasc Surg 2004;128(5):677–83.

[10] Januzzi JL, Isselbacher EM, Fattori R, Cooper JV, Smith DE, et al. Characterizing the young patient with aortic dissection: results from the International Registry of Aortic Dissection (IRAD). J Am Coll Cardiol 2004;43(4):665–9.

[11] Holmes KW, Maslen CL, Kindem M, Kroner BL, Song HK, Ravekes W, et al. GenTAC registry report: Gender differences among individuals with genetically triggered thoracic aortic aneurysm and dissection. Am J Med Genet A 2013;161(4):779–86.

[12] Baron M. An overview of the Notch signalling pathway. Semin Cell Dev Biol 2003;14(2):113–9.

[13] Fernández B, Durán AC, Fernández-Gallego T, Fernández MC, Such M, Arqué JM, et al. Bicuspid aortic valves with different spatial orientations of the leaflets are distinct etiological entities. J Am Coll Cardiol 2009;54(24):2312–8.

[14] Tzemos N, Therrien J, Thanassoulis G, Tremblay S, Jamorski MT, Webb GD, et al. Outcomes in adults with bicuspid aortic valves. J Am Coll Cardiol 2008;300(11):1317–25.

[15] Michelena HI, Desjardins VA, Avierinos J-F, Russo A, Nkomo VT, Sundt TM, et al. Natural history of asymptomatic patients with normally functioning or minimally dysfunctional bicuspid aortic valve in the community. Circulation 2008;117(21):2776–84.

[16] Verma S, Siu SC. Aortic dilatation in patients with bicuspid aortic valve. N Engl J Med 2014;370(20):1920–9.

[17] Duran AC, Frescura C, Sans-Coma V, Angelini A, Basso C, Thiene G. Bicuspid aortic valves in hearts with other congenital heart disease. J Heart Valve Dis 1995;4(6):581–90.

[18] Fernandes SM, Sanders SP, Khairy P, Jenkins KJ, Gauvreau K, Lang P, et al. Morphology of bicuspid aortic valve in children and adolescents. J Am Coll Cardiol 2004;44(8):1648–51.

[19] Ciotti GR, Vlahos AP, Silverman NH. Morphology and function of the bicuspid aortic valve with and without coarctation of the aorta in the young. Am J Cardiol 2006;98(8):1096–102.

[20] Brenner JI, Berg KA, Schneider DS, Clark EB, Boughman JA. Cardiac malformations in relatives of infants with hypoplastic left-heart syndrome. Am J Dis Child 1989;143(12): 1492–4.

[21] Hinton RB Jr, Martin LJ, Tabangin ME, Mazwi ML, Cripe LH, Benson DW. Hypoplastic left heart syndrome is heritable. J Am Coll Cardiol 2007;50(16):1590–5.

[22] Roberts WC, Morrow AG, Braunwald E. Complete interruption of the aortic arch. Circulation 1962;26:39–59.

[23] Bolling SF, Iannettoni MD, Dick M II, Rosenthal A, Bove EL. Shone's anomaly: operative results and late outcome. Ann Thorac Surg 1990;49(6):887–93.

[24] Topaz O, DeMarchena EJ, Perin E, Sommer LS, Mallon SM, Chahine RA. Anomalous coronary arteries: angiographic findings in 80 patients. Int J Cardiol 1992;34(2):129–38.

[25] Rashid A, Saucedo JF, Hennebry TA. Association of single coronary artery and congenital bicuspid aortic valve with review of literature. J Interv Cardiol 2005;18(5):389–91.

[26] Hutchins GM, Nazarian IH, Bulkley BH. Association of left dominant coronary arterial system with congenital bicuspid aortic valve. Am J Cardiol 1978;42(1):57–9.

[27] Hagan PG, Nienaber CA, Isselbacher EM, Bruckman D, Karavite DJ, Russman PL, et al. The International Registry of Acute Aortic Dissections (IRAD). JAMA-J Am Med Assoc 2000;283(7):897–903.

[28] Sundström J, Lind L, Arnlöv J, Zethelius B, Andrén B, Lithell HO. Echocardiographic and electrocardiographic diagnoses of left ventricular hypertrophy predict mortality independently of each other in a population of elderly men. Circulation 2001;103(19):2346–51.

[29] Talwar S, Downie PF, Squire IB, Davies JE, Barnett DB, Ng LL. Plasma N-terminal pro BNP and cardiotrophin-1 are elevated in aortic stenosis. Eur J Heart Fail 2001;3(1):15–9.

[30] Galan A, Zoghbi WA, Quiñones MA. Determination of severity of valvular aortic stenosis by Doppler echocardiography and relation of findings to clinical outcome and agreement with hemodynamic measurements determined at cardiac catheterization. Am J Cardiol 1991;67(11):1007–12.

[31] Baumgartner H, Hung J, Bermejo J, Chambers JB, Evangelista A, Griffin BP, et al. Echocardiographic assessment of valve stenosis: EAE/ASE recommendations for clinical practice. Eur J Echocardiogr 2009;10(1):1–25.

[32] Bhave NM, Lang RM. Evaluation of left ventricular structure and function by three-dimensional echocardiography. Curr Opin Crit Care 2013;19(5):387–96.

[33] Lancellotti P, Tribouilloy C, Hagendorff A, Moura L, Popescu BA, Agricola E, et al. European Association of Echocardiography recommendations for the assessment of valvular regurgitation. Part 1: aortic and pulmonary regurgitation (native valve disease). Eur J Echocardiogr 2010;11(3):223–44.

[34] Evangelista A, Flachskampf FA, Erbel R, Antonini-Canterin F, Vlachopoulos C, Rocchi G, et al. Echocardiography in aortic diseases: EAE recommendations for clinical practice. Eur J Echocardiogr 2010;11(8):645–58.

[35] Hiratzka LF, Bakris GL, Beckman JA, Bersin RM, Carr VF, Casey DE Jr, et al. 2010 ACCF/AHA/AATS/ACR/ASA/SCA/SCAI/SIR/STS/SVM guidelines for the diagnosis and management of patients with Thoracic Aortic Disease: a report of the American College of Cardiology Foundation/American Heart Association Task Force on Practice Guidelines, American Association for Thoracic Surgery, American College of Radiology, American Stroke Association, Society of Cardiovascular Anesthesiologists, Society for Cardiovascular Angiography and Interventions, Society of Interventional Radiology, Society of Thoracic Surgeons, and Society for Vascular Medicine. Circulation 2010;121(13):e266–369.

[36] Wanamaker KM, Amadi CC, Mueller JS, Moraca RJ. Incidence of aortic arch anomalies in patients with thoracic aortic dissections. J Cardiac Surg 2013;28(412):151–4.

[37] Lobato AC, Puech-Leão P. Predictive factors for rupture of thoracoabdominal aortic aneurysm. J Vasc Surg 1998;27(3):446–53.

[38] Beaton AZ, Nguyen T, Lai WW, Chatterjee S, Ramaswamy P, Lytrivi ID, et al. Relation of coarctation of the aorta to the occurrence of ascending aortic dilation in children and young adults with bicuspid aortic valves. Am J Cardiol 2009;103(2):266–70.

[39] Oliver JM, Alonso-Gonzalez R, Gonzalez AE, Gallego P, Sanchez-Recalde A, Cuesta E, et al. Risk of aortic root or ascending aorta complications in patients with bicuspid aortic valve with and without coarctation of the aorta. Am J Cardiol 2009;104(7):1001–6.

[40] Nishimura RA, Otto CM, Bonow RO, Carabello BA, Erwin JP III, Guyton RA, et al. 2014 AHA/ACC guideline for the management of patients with valvular heart disease: A report of the American College of Cardiology/American Heart Association Task Force on Practice Guidelines. J Am Coll Cardiol 2014;63(22):1001–6.

[41] Vahanian A, Alfieri O, Andreotti F, Antunes M. Guidelines on the management of valvular heart disease. Eur Heart J 2012;33(19):2451–96.

[42] Cripe L, Andelfinger G, Martin LJ, Shooner K, Benson DW. Bicuspid aortic valve is heritable. J Am Coll Cardiol 2004;44(1):138–43.

[43] Erbel R, Aboyans V, Boileau C, Di Bartolomeo R, Eggebrecht H, Frank H, et al. 2014 ESC Guidelines on the diagnosis and treatment of aortic diseases: document covering acute and chronic aortic diseases of the thoracic and abdominal aorta of the adult. The Task Force for the Diagnosis and Treatment of Aortic Diseases of the European Society of Cardiology (ESC). Eur Heart J 2014;35(41):2873–926.

[44] Regitz-Zagrosek V, Blomstrom Lundqvist C, Borghi C, Cifkova R, Ferreira R, Foidart JM, et al. ESC Guidelines on the management of cardiovascular diseases during pregnancy: the Task Force on the Management of Cardiovascular Diseases during Pregnancy of the European Society of Cardiology (ESC). Eur Heart J 2011;32(24):3147–97.

Chapter 4e

Turner Syndrome

A.T. van den Hoven, BSc, J.W. Roos-Hesselink, MD, PhD, J. Timmermans, MD

1 INTRODUCTION

Turner syndrome, a partial or complete monosomy of the X chromosome, is a genetic disorder that occurs in 1 per 2500 live-born females and was originally described by Henry Turner in 1938 [1,2]. Patients may suffer from a multitude of disorders, including short stature, estrogen deficiency, infertility, and a "webbed neck" [3]. Morbidity and mortality rates are significantly higher in Turner patients [1,4]. Turner patients usually receive care from a multidisciplinary team in a tertiary center; such a team often comprises a pediatrician, a gynecologist, an endocrinologist-internist, and a cardiologist. For complex cases, it may also be necessary to involve ear-nose-throat specialists, clinical-geneticists, ophthalmologists, psychologists, orthodontologists, and orthopedic surgeons. Recently, the cardiovascular aspect of the syndrome has received more attention; according to current guidelines, every patient is advised to visit a cardiologist who specializes in congenital cardiology at least every 5 years [5]. Due to increasingly complex patient care, it is important that all patient care providers are aware of the cardiovascular phenotype associated with Turner syndrome.

2 GENETICS

Females typically have two X chromosomes: one paternally derived (X^p) and one maternally derived (X^m). However, in patients with Turner syndrome, a de novo nondisjunction of the X chromosome can lead to a female with a completely or partially absent X chromosome, most often the one that is paternally derived. This nondisjunction results in a diverse spectrum of karyotypes, of which the nonmosaic 45,X0 monosomy is the most well known. Other known karyotypes associated with Turner syndrome are different forms of mosaicism (eg, 45X/46XX) and structurally abnormal X chromosomes, such as iso-chromosomes (eg, 46X,i(Xq)); ring chromosomes (eg, 46X,r(X)); deletions (eg, 46,X,del(X)); and even karyotypes with Y-chromosomal DNA (eg, 45,X/46XY).

Aneurysms-Osteoarthritis Syndrome. http://dx.doi.org/10.1016/B978-0-12-802708-0.00010-7
Copyright © 2017 Elsevier Inc. All rights reserved.

The suggestion that the nondisjunction in Turner syndrome is the result of meiotic factors is unlikely, because the number of 45,X conceptions is too high to be explained solely by the frequency of gametes hypohaploid for a sex chromosome [6]. A loss of sex chromosome after conception (a mitotic loss) would better explain the unequal ratio of the parental origin of the X chromosome (male-to-female is 1:3). A 46,XX conception, upon loss of one X chromosome during mitosis, would generate a 45,X line with equal paternal and maternal origin of the remaining X chromosome; whereas a 46,XY conception would generate a 45,X cell line of maternal origin. This explains the 1:3 male-to-female ratio.

An estimated 1 in every 100 pregnancies starts as a Turner syndrome (45,X0) pregnancy. However, 99% of these pregnancies do not make it to full term [6]. In about 50% of the cases, analysis of peripheral lymphocytes indicates the complete loss of one X chromosome, most often the paternal X chromosome. However, most studies have an inherent bias, because 45,X0 is overrepresented in clinical populations; these patients are more prone to displaying the Turner phenotype. The frequency of the chromosomal pattern varies, depending on the reason for karyotyping [7]. Karyotype determination that is carried out because of prenatal echo findings shows a 45,X0 karyotype in 90% of cases, whereas it is only 63% in accidental findings [5].

In addition, with the use of more sensitive genetic techniques, such as fluorescence in situ hybridization and reverse transcriptase polymerase-chain-reaction assays, nonmosaic 45,X0 prevalence rises to 60% and 74%, respectively. This may suggest that the survival of nonmosaic karyotypes is an even rarer event than previously assumed. Thus, the hypothesis by Ernest Hook and Dorothy Warburton that all Turner syndrome females might actually be "cryptic mosaics" has gained ground.

However, the question remains regarding why the partial or whole absence of an X chromosome should be so invalidating when approximately 50% of the world population seems to do fine with one X chromosome. Two main theories exist that try to explain the phenotypes found in Turner syndrome; a visual depiction of both theories is provided in Fig. 4e.1.

For the first theory, called the X-inactivation theory, it is important to understand the concept of X-inactivation, which occurs when fully transcribed somatic cells in 46,XX females would result in a surplus of transcriptional product. Therefore, one X chromosome has to be transcriptionally silenced to effectively reduce the transcriptional product to that of males. However, approximately 25% of the genes on the inactive X chromosome escape silencing [8]. These genes are predominantly located in two regions, called the pseudo-autosomal regions 1 and 2 (PAR1 and PAR2). PAR1 (2.6 Mbp) is located on the p-arms end of both the X and Y chromosomes, and PAR2 (320 kbp) is located on the q-arms end. Pseudo-autosomal regions contain genes that normally escape X-inactivation have a Y-chromosome homolog and are inherited like autosomal genes. The X-inactivation theory postulates that the haploinsufficiency of these

(A) Imprinting (B) X inactivation

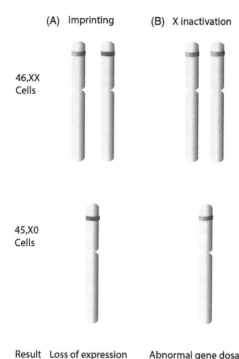

46,XX
Cells

45,X0
Cells

Result Loss of expression Abnormal gene dosage

FIGURE 4E.1 **The parental imprinting theory (A) and X-inactivation theory (B).**

pseudo-autosomal genes on the X chromosome results in an insufficient dosage to ensure normal expression [9]. Which exact genes contribute to the Turner phenotype is not yet clear, but the diverse spectrum of features in patients with Turner syndrome suggests that multiple genes may contribute. Genes located in PAR1 have already been identified; the short-stature-homeobox (*SHOX*) gene is one example. Haploinsufficiency for this gene, which escapes X-inactivation, appears to cause the short stature in patients with Turner syndrome. More recently, a study by Achia Urbach and Nissim Benvenitsy discovered a gene necessary for placental function (*PSF2RA*) that was located in the pseudo-autosomal region [10]. According to this study, deletion of this gene may cause placental malfunction, leading to high rates of fetal mortality in nonmosaic Turner syndrome.

The second theory suggests that Turner syndrome might be the effect of imprinted genes, expressed in a monoallelic fashion, depending on the parental origin cause [11]. Parental imprinting is a form of epigenetic regulation that results in parent-of-origin differential gene expression [12]. It has a crucial role in prenatal growth and placentation and affects the development of the musculoskeletal system and the brain. Hence, as displayed in Fig. 4e.1, there is no expression when the X chromosome containing the expressed allele is lost. The loss of these active alleles has also been implicated in Prader-Willi syndrome [13]. In nonmosaic Turner syndrome, it is often the maternal X chromosome that is retained [14]. Phenotypical traits in Turner syndrome have been associated with the parental origin of the remaining X chromosome.

In conclusion, the genetic aspects of Turner syndrome have not yet been fully unraveled. The roles of imprinting and X-inactivation must be investigated further.

3 CARDIOVASCULAR DISEASE

An estimated 50% of women with Turner syndrome suffer from cardiovascular disease, which can be congenital or acquired [15,16]. Various congenital abnormalities complicate care and are likely to cause significant morbidity. The congenital heart defects are mainly left-sided, of which a bicuspid aortic valve (BAV) (15–30%), elongation of the transverse aortic arch (49%), and coarctation of the aorta (17%) are most prevalent [8,16]. Associated venous lesions include partial abnormal pulmonary venous return and persistent left superior vena cava [16,17]. Other defects, such as a ventricular septal defect, common in Down syndrome, are seen less often in Turner syndrome [18]. The acquired heart diseases mainly include hypertension, aortic dilatation, and dissection [19,20].

3.1 Etiology

Genetic disorders, such as Turner syndrome, are sometimes first identified by a prenatal ultrasound after the detection of nuchal translucency. This collection of fluid under the skin in the neck region is often a precursor of the webbed neck, one of the classic traits associated with Turner syndrome. Among infants with webbed necks, 68% are affected by a genetic syndrome, such as Down (37%), Noonan (5%), or Turner (13%) [21]. Congenital heart defects are detected in 60% of patients with webbed necks, and in patients with Turner syndrome, a high prevalence of cardiac abnormalities, such as aortic coarctation, has also been observed [18,21–23].

Therefore, a common causal mechanism from which the multiple congenital heart defects arise has been suggested to lie in the disturbance of early lymphangiogenesis [24]. This co-occurrence does not necessarily imply a causal relation, because a single gene causing both lymphedema and congenital heart defects could theoretically confound this causal relation [23]. Indeed, haploinsufficiency for the autosomal gene *FOXC2;16q* seems to cause lymphedema and cardiac defects independent of each other [22]. An attempt to further elucidate a causality, genetic or otherwise, between lymphangiogenesis and congenital heart defects must start with a thorough understanding of the cardiovascular phenotype.

3.2 Congenital Abnormalities

As the accuracy of imaging modalities and genetic techniques increases, a growing number of congenital cardiovascular abnormalities are being detected in patients with Turner syndrome, and genotype–phenotype correlations may become more evident.

3.2.1   Bicuspid Aortic Valve

The earliest description of the BAV dates back to the 15th century, when Leonardo da Vinci sketched different variants of the aortic valve [25]. Furthermore, the association with aortic regurgitation and stenosis has been known for 150 years [26]. With a prevalence in the general population of about 0.5–2% (males-to-females is 3:1), it is the most common congenital heart defect [27]. Approximately one-third of these patients will develop serious complications that require treatment, because a BAV is likely to become stenotic or insufficient [28,29]. From a developmental viewpoint, the BAV is thought to be more than the mere fusion of two cusp leaflets; it is seen as a part of a developmental defect ranging from uni- to quadricuspid valves [6]. A true bicuspid valve is very rarely seen. More often, it is a fusion of two cusps, resulting in two remaining cusps that are often unequal in size due to the fusion. In this chapter, we focus on BAV patients who have Turner syndrome, because the bicuspid valve in the general population is described in Chapter 4d.

The prevalence of a BAV in patients with Turner syndrome ranges quite spectacularly, from as low as 10% to as high as 39%, depending on the imaging modality and study population [30]. Autosomal-dominant, X-linked, and familial modes of inheritance have been reported in the general population [27]. BAVs are associated with a monosomy 45,X0 and are often seen in combination with a coarctation of the aorta, but they are also associated with acquired diseases, such as aortic dilatation, aneurysm, and dissection [31]. Two-dimensional and Doppler echocardiography is currently the most widely used and least-demanding technique to assess valvular function. However, some studies suggest it underestimates the prevalence of BAV when compared to magnetic resonance imaging (MRI) [22]. Valves that are hard to assess seem to be bicuspid more often, resulting in an underestimation by echocardiography technicians [22].

Treatment for BAV in patients with Turner syndrome does not differ from the norm, but it is important to note that the aorta of BAV patients dilates more quickly than does the tricuspid aortic valve in the general population, especially at the level of the ascendens and sinus [17]. Aortic dilatation increases the risk of aortic dissection, but aortic dissection may also occur at normal aortic diameters; it is therefore advisable to pay extra attention to Turner syndrome patients when they display symptoms [19]. The risk of dissection is discussed further below.

3.2.2   Coarctation of the Aorta

Aortic coarctation is a congenital narrowing of the aorta, distal to the aortic annulus, that occurs in 3.4 per 10,000 live births and constitutes 5% of all congenital heart defects in the general population [32]. Approximately 17% of Turner syndrome patients have a coarctation of the aorta [8]. Moreover, aortic coarctation in Turner syndrome patients is often associated with the presence of a BAV (RR = 4.6) [22,33]. Vincent Ho et al. recently found that aortic coarctation appeared to be associated with an elongated transverse aortic arch [16].

The first successful surgical correction of an aortic coarctation was performed in 1945 [34]. Nowadays, aortic coarctation is still preferably repaired surgically at an early age. However, lifelong surveillance is a necessity, because patients remain at risk for recoarctation and aneurysm formation [35–37]. Stent implantation was introduced as a treatment for aortic coarctation in the late 1980s, providing good gradient relief and a low complication rate [38–40].

Elliot Shinebourne and A.M. Elseed hypothesized a hemodynamic pathogenesis of aortic coarctation [41]. The altered flow patterns could be caused by a left-sided blockage within the fetal circulation, resulting in elevated pulmonary pressure and blood flow over the ductus arteriosus. They predicted that abnormal flow via the ductus to the isthmic portion of the arch would produce hypoplasia, tortuosity, or coarctation of the aorta in the juxtaductal region. These are all abnormalities that also appear in patients with Turner syndrome, some of which are predicted to be the result of left-sided lymphatic compression of the aortic arch, as hypothesized by Edward Clark [24]. These abnormalities also include some right-sided defects (eg, partial abnormal pulmonary venous return, persistent left superior vena cava), due to back pressure. However, contributions from genetic regulatory mechanisms to these malformations cannot be ruled out [8].

More recently, the aortic arch and aortic wall composition in patients with Turner syndrome have received increasing attention. Changes in vascular smooth muscle cells, elastin, and collagen fiber appear to contribute to their cardiovascular problems [42]. Cystic medial wall necrosis, similar to what can be found in patients with Marfan syndrome, has been described in patients with Turner syndrome and is suggested to be a causative factor of aortic dissection [43–45]. The aortic wall certainly appears to be fragile in Turner syndrome patients as well. Stent implantation may therefore be associated with a higher risk of complications, especially aortic dissection. However, data on the ideal corrective technique of coarctation repair for patients with Turner syndrome are limited, because it is largely based on small case series or case reports, the results of which are often contradictory [46,47].

3.2.3 Congenital Aortic Arch Abnormalities

3.2.3.1 Aortic Arch

Recently an elongated transverse arch has been added to the Turner syndrome cardiac phenotype. It is defined by two criteria: (1) an origin of the left subclavian artery and (2) an inward indentation of the lesser curvature or kinking at the aortic isthmus [16]. It has been reported to occur in approximately half of the patients and is associated with higher blood pressure, aortic coarctation, aberrant right subclavian artery, and a left superior vena cava.

3.2.3.1.1 Aberrant right subclavian artery The aberrant right subclavian artery, or arteria lusoria, is the most common anomaly of the aortic arch, which may occur in 0.4–2% of the population [48]. For women with Turner syndrome, it can occur in as many as 8%. Its clinical significance lies in the fact that it can cause dysphagia and mask the presence of a coarctation by altering the upper to lower blood pressure ratio, when measured at the right upper extremity [16,49]. Little is known about the aortic branching pattern in Turner syndrome and its relation with other congenital heart defects seen in the cardiovascular phenotype. ***3.2.3.1.2 Bovine arch*** A common origin of the innominate artery and the left common carotid, also known as a "bovine aortic arch," is seen in 8% of women with Turner syndrome but has not yet been officially correlated with the syndrome, because it is also observed in 13% of the general population [16].

3.2.4 Venous Anomalies

In patients with Turner syndrome, the cardiac defects are often left-sided, and not many venous abnormalities are associated with the syndrome. As stated before, with more advanced imaging techniques, the anatomy can be mapped in more detail, leading to the discovery of rarer cardiovascular malformations.

3.2.4.1 Partial Abnormal Pulmonary Venous Return

Partial anomalous pulmonary venous return, first described by Winslow in 1739, is often found by chance during routine check-up and can cause a hemodynamically significant left-to-right shunt [50]. Significant shunts (Qp:Qs > 1.5:1.0) can manifest as right heart volume overload, or the onset of pulmonary hypertension, and can eventually result in right ventricular hypertrophy or failure [51,52]. Therefore, this condition necessitates early diagnosis and treatment. The prevalence of partial abnormal pulmonary venous return in patients with Turner syndrome might be underestimated, because it is difficult to diagnose via echocardiography. Previously, venous abnormalities in patients with Turner syndrome were relatively unknown, and their occurrence was grossly underestimated due to this inadequate method of diagnosis. In a study using echocardiography, Daniela Prandstraller et al. reported a partial abnormal pulmonary venous return prevalence of 2.9%; however, since MRI and computed tomography came in to regular use over the last 5 years, the prevalence of partial abnormal pulmonary venous return in patients with Turner syndrome has been suggested to be as high as 16% by Ho et al. [16,17,53]. Recently, more detailed data showed a partial abnormal pulmonary venous return in almost 25% of Turner syndrome patients.

3.2.4.2 Persistent Left-Sided Superior Vena Cava

A persistent left-sided superior vena cava is seen in 0.3–0.5% of the normal population and in 4.4% of those with congenital heart defects [54,55]. Most often, it is seen incidentally during a computed tomography scan of the thorax. In addition to the persistent left-sided superior vena cava (82–90%), a normal vena cava can also be found [55]. Left to right shunting can be present, as the vein drains into the left atrium in 8% of cases, but it is often not clinically significant. During fetal development, the left anterior cardinal vein normally disintegrates, but in some cases, this fails to take place. Failure of the left anterior cardinal vein to disintegrate results in connections with either the coronary sinus (92%) or the left atrium (8%) [55].

3.2.4.3 Interrupted Inferior Vena Cava with Azygous Continuation

This venous malformality has been described anecdotally in case reports and has been associated with congenital heart defects in the past [56]. Very few cases are known in literature, but it can present as a dilated azygos vein or with pulmonary hypertension [57]. A missed diagnosis can lead to problems during surgical procedures or percutaneous interventions. However, it is unclear whether any causal relation is present, because the prevalence in the general population is not known. Larger cohorts with a control population will have to specifically be examined for this defect before we can draw any conclusions regarding its link with Turner syndrome.

3.3 Acquired Heart Disease

Acquired heart disease is a significant cause of morbidity and mortality in patients with Turner syndrome, as was revealed in an article by Mortensen et al [8]. Aortic dilation and dissection are largely to blame for the absolute excess mortality among the Turner syndrome population (standardized mortality ratio = 23.6). Ischemic heart disease (standardized mortality ratio = 2.8) is also a significant contributor to mortality in Turner syndrome patients, especially at older ages [8].

3.3.1 Aortic Dissection

Acute aortic dissection often presents with a sharp pain, but its clinical presentation is frequently more diverse [58]. The incidence is estimated at 36 per 100,000 patient–years, compared to 6 per 100,000 patient–years in the general population (male-to-female ratio is 2:1) [20]. Dissection also occurs much earlier than in the general population, with 56% of dissections between the age of 20 and 40 years, an incidence of 14 per 100,000 before age 19, and an average dissection age of 35 years [20]. Risk factors for dissection in patients with Turner syndrome include hypertension; karyotype 45,X0; BAV; aortic coarctation; age; and pregnancy [8,59]. However, it remains unclear whether this high rate of dissection can occur separately from the aforementioned risk factors, as it does in connective tissue disorders, such as Marfan, Loeys-Dietz, or aneurysms-osteoarthritis syndrome. Some articles do suggest that Turner syndrome is a separate risk factor for aortic dilatation [59].

There is no data on the outcome of dissection in patients with Turner syndrome, but there is no reason to suspect it to be less severe than in the general population, where mortality rates vary depending on the type: 26% for type A dissection and 10.7% for type B dissection, respectively [58].

3.3.2 Aortic Dilatation

Aortic dilatation is estimated to occur in up to 42% of patients with Turner syndrome [15,18,43,60]. Body size and age are the primary determinants of aortic size in patients with Turner syndrome, and because they are generally smaller and have "barrel-shaped" chests, it is important to correct their aortic dimensions for body surface area [19]. Several factors, such as BAV, hypertension, and vessel wall structure, contribute to aortic dilatation. The presence of a BAV is associated with dilatation of the aortic root and proximal ascending aorta, which can be attributed to either changes in flow or to abnormalities of the aortic media. Recent studies have shown a clear role for cellular mechanisms underlying the dilatation and its prevalence in first-degree relatives of BAV patients [61].

Therefore, it is particularly important to closely monitor aortic dimensions in females with Turner syndrome. Annual intensive follow-up may be justified when absolute ascending aortic diameters exceed 40 mm or 2.1 cm/m^2, and early surgical intervention might be necessary in this population [27]. Dutch guidelines advise using MRI to monitor aortic diameters in these patients and determining the correct therapy by referring to the aortic size index [5]. The guidelines also state that frequent follow-up (every 1–2 years) is justified when aortic size index > 2.0 cm/m^2. It is also advisable to consider medicinal treatment with beta-blockers and angiotensin receptor blockers to control blood pressure. Elective surgery might even be considered when the aortic size index exceeds 2.5 cm/m^2 or when rapid progression of the aortic diameter (0.5 cm/year) is observed.

3.3.3 Hypertension

Aortic root dilatation is closely associated with blood pressure and left ventricular thickness and valve type, but it does not seem to be affected by atherosclerosis [60]. Hypertension occurs in 7–17% of young girls with Turner syndrome; in 50% of young adults, it can be secondary to an aortic coarctation or kidney disease but is often primary [62]. Because it is a risk factor for dissection, guidelines advise taking blood pressure measurements once or twice per year and striving for a target blood pressure of < 140 mmHg, or in case of a bicuspid valve, < 120 mmHg [5]. Hormone substitution therapy appears to positively influence blood pressure, or at least have no negative influence [62].

3.4 Pregnancy

Infertility is one of the important complications of Turner syndrome affecting a woman's life. Patients with some mosaic karyotypes (45,X/46,XX) may be able to achieve spontaneous pregnancies (2–6%), while others will only be able to conceive by oocyte donation [63]. However, these assisted reproductive technologies may increase the risk of adverse events in Turner syndrome patients, such as aortic dissection or rupture [64]. This risk seems to be augmented by hormonal influences on the vascular wall [64,65]. Maternal death from aortic dissection in Turner syndrome pregnancies is estimated at 2%, a 100-fold increased risk as compared to the general population [65,66]. The presence of hypertension, BAV, and aortic coarctation is associated with an increased risk, and pregnancy itself seems to be an additional, separate risk factor [67]. Therefore, treatment of hypertension, which is associated with poor fetal outcomes, such as prematurity and fetal growth retardation [68], is of great importance for women and their children.

Special attention should also be given to the aortic diameter, as an aortic size index > 2 cm/m^2 and/or a significant abnormality is a strict contraindication for attempting pregnancy in Turner syndrome patients [65]. Aortic diameters should be measured at least once every 4–8 weeks [69]. Consequently, deliveries should take place in a medical center with cardiothoracic surgery facilities readily available.

REFERENCES

[1] Stochholm K, et al. Prevalence, incidence, diagnostic delay, and mortality in Turner syndrome. J Clin Endocr Metab 2006;91(10):3897–902.

[2] Turner H. A syndrome of infantilism, congenital webbed neck, and cubitus valgus. Endocrinology 1938;23:556–74.

[3] Bondy CA, et al. The physical phenotype of girls and women with Turner syndrome is not X-imprinted. Hum Genet 2007;121(3–4):469–74.

[4] Gravholt CH, et al. Morbidity in Turner syndrome. J Clin Epidemiol 1998;51(2):147–58.

[5] van den Akker EL, van Alfen AA, Sas T, Kerstens M, Cools M, Lambalk CB. Clinical guideline Turner Syndrome. Ned Tijdschr Geneeskd 2014;158:A7375.

[6] Hook EB, Warburton D. Turner syndrome revisited: review of new data supports the hypothesis that all viable 45,X cases are cryptic mosaics with a rescue cell line, implying an origin by mitotic loss. Hum Genet 2014;133(4):417–24.

[7] Gunther DF, et al. Ascertainment bias in Turner syndrome: new insights from girls who were diagnosed incidentally in prenatal life. Pediatrics 2004;114(3):640–4.

[8] Mortensen KH, Andersen NH, Gravholt CH. Cardiovascular phenotype in Turner syndrome—integrating cardiology, genetics, and endocrinology. Endocr Rev 2012;33(5):677–714.

[9] Lyon MF. Gene action in the X-chromosome of the mouse (Mus musculus L.). Nature 1961;190:372–3.

[10] Urbach A, Benvenisty N. Studying early lethality of 45,XO (Turner's syndrome) embryos using human embryonic stem cells. PLoS One 2009;4(1):e4175.

[11] Cattanach BM, Kirk M. Differential activity of maternally and paternally derived chromosome regions in mice. Nature 1985;315(6019):496–8.

[12] Stelzer Y, et al. The noncoding RNA IPW regulates the imprinted DLK1-DIO3 locus in an induced pluripotent stem cell model of Prader-Willi syndrome. Nat Genet 2014;46(6):551–7.

[13] Nicholls RD, et al. Genetic imprinting suggested by maternal heterodisomy in nondeletion Prader-Willi syndrome. Nature 1989;342(6247):281–5.

[14] Sagi L, et al. Clinical significance of the parental origin of the X chromosome in Turner syndrome. J Clin Endocr Metab 2007;92(3):846–52.

[15] Dawson-Falk KL, et al. Cardiovascular evaluation in Turner syndrome: utility of MR imaging. Australas Radiol 1992;36(3):204–9.

[16] Ho VB, et al. Major vascular anomalies in Turner syndrome: prevalence and magnetic resonance angiographic features. Circulation 2004;110(12):1694–700.

[17] Kim HK, et al. Cardiovascular anomalies in Turner syndrome: spectrum, prevalence, and cardiac MRI findings in a pediatric and young adult population. Am J Roentgenol 2011;196(2):454–60.

[18] Mazzanti L, Cacciari E. Congenital heart disease in patients with Turner's syndrome. Italian Study Group for Turner Syndrome (ISGTS). J Pediatr 1998;133(5):688–92.

[19] Matura LA, et al. Aortic dilatation and dissection in Turner syndrome. Circulation 2007;116(15):1663–70.

[20] Gravholt CH, et al. Clinical and epidemiological description of aortic dissection in Turner's syndrome. Cardiol Young 2006;16(5):430–6.

[21] Berdahl LD, Wenstrom KD, Hanson JW. Web neck anomaly and its association with congenital heart disease. Am J Med Genet 1995;56(3):304–7.

[22] Sachdev V, et al. Aortic valve disease in Turner syndrome. J Am Coll Cardiol 2008;51(19):1904–9.

[23] Loscalzo ML, et al. Association between fetal lymphedema and congenital cardiovascular defects in Turner syndrome. Pediatrics 2005;115(3):732–5.

[24] Clark EB. Neck web and congenital heart defects: a pathogenic association in 45 X-O Turner syndrome? Teratology 1984;29(3):355–61.

[25] Braverman AC, et al. The bicuspid aortic valve. Curr Prol Cardiology 2005;30(9):470–522.

[26] Peacock TB. Valvular disease of the heart. London: Churchill; 1865. p. 2–33.

[27] Siu SC, Silversides CK. Bicuspid aortic valve disease. J Am Coll Cardiol 2010;55(25):2789–800.

[28] Michelena HI, et al. Natural history of asymptomatic patients with normally functioning or minimally dysfunctional bicuspid aortic valve in the community. Circulation 2008;117(21):2776–84.

[29] Friedman T, Mani A, Elefteriades JA. Bicuspid aortic valve: clinical approach and scientific review of a common clinical entity. Expert Rev Cardiovasc Ther 2008;6(2):235–48.

[30] Gutmark-Little I, Backeljauw PF. Cardiac magnetic resonance imaging in Turner syndrome. Clin Endocrinol 2013;78(5):646–58.

[31] Lindsay J Jr. Coarctation of the aorta, bicuspid aortic valve and abnormal ascending aortic wall. Am J Cardiol 1988;61(1):182–4.

[32] van der Linde D, et al. Birth prevalence of congenital heart disease worldwide: a systematic review and meta-analysis. J Am Coll Cardiol 2011;58(21):2241–7.

[33] Mortensen KH, et al. Abnormalities of the major intrathoracic arteries in Turner syndrome as revealed by magnetic resonance imaging. Cardiol Young 2010;20(2):191–200.

[34] Crafoord C. Congenital coarctation of the aorta and its surgical treatment. J Thorac Surg 1945;14:347–61.

[35] Toro-Salazar OH, et al. Long-term follow-up of patients after coarctation of the aorta repair. Am J Cardiol 2002;89(5):541–7.

[36] Cohen M, et al. Coarctation of the aorta. Long-term follow-up and prediction of outcome after surgical correction. Circulation 1989;80(4):840–5.

[37] Brown ML, et al. Coarctation of the aorta: lifelong surveillance is mandatory following surgical repair. J Am Coll Cardiol 2013;62(11):1020–5.

[38] Chessa M, et al. Results and mid-long-term follow-up of stent implantation for native and recurrent coarctation of the aorta. Eur Heart J 2005;26(24):2728–32.

[39] Hamdan MA, et al. Endovascular stents for coarctation of the aorta: initial results and intermediate-term follow-up. J Am Coll Cardiol 2001;38(5):1518–23.

[40] Magee AG, et al. Stent implantation for aortic coarctation and recoarctation. Heart 1999;82(5):600–6.

[41] Shinebourne EA, Elseed AM. Relation between fetal flow patterns, coarctation of the aorta, and pulmonary blood flow. Br Heart J 1974;36(5):492–8.

[42] El-Hamamsy I, Yacoub MH. Cellular and molecular mechanisms of thoracic aortic aneurysms. Nat Rev Cardiol 2009;6(12):771–86.

[43] Lin AE, et al. Aortic dilation, dissection, and rupture in patients with Turner syndrome. J Pediatr 1986;109(5):820–6.

[44] Price WH, Wilson J. Dissection of the aorta in Turner's syndrome. J Med Genet 1983;20(1):61–3.

[45] Kostich ND, Opitz JM. Ullrich-Turner syndrome associated with cystic medial necrosis of the aorta and great vessels: case report and review of the literature. Am J Med 1965;38:943–50.

[46] Zanjani KS, et al. Usefulness of stenting in aortic coarctation in patients with the Turner syndrome. Am J Cardiol 2010;106(9):1327–31.

[47] Fejzic Z, van Oort A. Fatal dissection of the descending aorta after implantation of a stent in a 19-year-old female with Turner's syndrome. Cardiol Young 2005;15(5):529–31.

[48] Azam AF, et al. Surgery for isolated non-inflammatory chronic total occlusion of the left main coronary artery: a case report and literature review. Med J Malaysia 2011;66(4):374–5.

[49] Asherson N. David Bayford. His syndrome and sign of dysphagia lusoria. Ann Roy Coll Surg 1979;61(1):63–7.

[50] Ammash NM, et al. Partial anomalous pulmonary venous connection: diagnosis by transesophageal echocardiography. J Am Coll Cardiol 1997;29(6):1351–8.

[51] Sears EH, Aliotta JM, Klinger JR. Partial anomalous pulmonary venous return presenting with adult-onset pulmonary hypertension. Pulm Circ 2012;2(2):250–5.

[52] Reid JR, editor. Pediatric Radiology. New York: Oxford University Press; 2013.

[53] Prandstraller D, et al. Turner's syndrome: cardiologic profile according to the different chromosomal patterns and long-term clinical follow-up of 136 nonpreselected patients. Pediatr Cardiol 1999;20(2):108–12.

[54] Kellman GM, et al. Computed tomography of vena caval anomalies with embryologic correlation. Radiographics 1988;8(3):533–56.

[55] Pretorius PM, Gleeson FV. Case 74: right-sided superior vena cava draining into left atrium in a patient with persistent left-sided superior vena cava. Radiology 2004;232(3):730–4.

[56] Bartram U, Fischer G, Kramer HH. Congenitally interrupted inferior vena cava without other features of the heterotaxy syndrome: report of five cases and characterization of a rare entity. Pediatr Devel Pathol 2008;11(4):266–73.

[57] Mehta AJ, et al. Interrupted inferior vena cava syndrome. J Assoc Physician India 2012;60: 48–50.

[58] Hagan PG, et al. The International Registry of Acute Aortic Dissection (IRAD): new insights into an old disease. JAMA-J Am Med Assoc 2000;283(7):897–903.

[59] Lopez L, et al. Turner syndrome is an independent risk factor for aortic dilation in the young. Pediatrics 2008;121(6):e1622–7.

[60] Elsheikh M, et al. Hypertension is a major risk factor for aortic root dilatation in women with Turner's syndrome. Clin Endocrinol 2001;54(1):69–73.

[61] Biner S, et al. Aortopathy is prevalent in relatives of bicuspid aortic valve patients. J Am Coll Cardiol 2009;53(24):2288–95.

[62] Gravholt CH. Epidemiological, endocrine and metabolic features in Turner syndrome. Arq Bras Endocrinol 2005;49(1):145–56.

[63] Bryman I, et al. Pregnancy rate and outcome in Swedish women with Turner syndrome. Fertil Steril 2011;95(8):2507–10.

[64] Karnis MF, et al. Risk of death in pregnancy achieved through oocyte donation in patients with Turner syndrome: a national survey. Fertil Steril 2003;80(3):498–501.

[65] Practice Committee of American Society for Reproductive Medicine. Increased maternal cardiovascular mortality associated with pregnancy in women with Turner syndrome. Fertil Steril 2012;97(2):282–4.

[66] Hadnott TN, et al. Outcomes of spontaneous and assisted pregnancies in Turner syndrome: the U.S. National Institutes of Health experience. Fertil Steril 2011;95(7):2251–6.

[67] Chevalier N, et al. Materno-fetal cardiovascular complications in Turner syndrome after oocyte donation: insufficient prepregnancy screening and pregnancy follow-up are associated with poor outcome. J Clin Endocr Metab 2011;96(2):E260–7.

[68] Bodri D, et al. Oocyte donation in patients with Turner's syndrome: a successful technique but with an accompanying high risk of hypertensive disorders during pregnancy. Hum Reprod 2006;21(3):829–32.

[69] Hiratzka LF, et al. 2010 ACCF/AHA/AATS/ACR/ASA/SCA/SCAI/SIR/STS/SVM guidelines for the diagnosis and management of patients with thoracic aortic disease: executive summary. A report of the American College of Cardiology Foundation/American Heart Association Task Force on Practice Guidelines, American Association for Thoracic Surgery, American College of Radiology, American Stroke Association, Society of Cardiovascular Anesthesiologists, Society for Cardiovascular Angiography and Interventions, Society of Interventional Radiology, Society of Thoracic Surgeons, and Society for Vascular Medicine. Catheter Cardio Inte 2010;76(2):E43–86.

Chapter 5

Cardiovascular Imaging in Aneurysm-Osteoarthritis Syndrome

R.G. Chelu, MD, D. van der Linde, MD, MSc, PhD, K. Nieman, MD, PhD

1 AORTIC ANATOMY

The aorta is a large conduit vessel transferring oxygenated blood from the left heart into the body. It is composed of three layers: the intima, the media, and the adventitia. The intima is composed of a monolayered endothelium; the media is the thickest layer and is composed of concentric layers of elastic tissue, smooth muscle cells, and collagen; and the adventitia is the external layer that contains the vasa vasorum that supplies blood to the aortic wall.

The anatomical segmentation of the aorta is important, because these segments are analyzed separately in the context of diagnosing aortic disease. The proximal part of the aorta, the aortic root, contains the aortic valve and the sinus of Valsalva. The next segment is the sinotubular junction, which is a well-defined border between the curved-shaped sinus of Valsalva and the ascending aorta. The ascending aorta extends between the sinotubular junction and the first arch vessel, usually the innominate artery. This aortic segment is located within the pericardium, as the pericardium reflects at the first arch vessel. The aortic arch is the transverse section of the thoracic aorta within the upper mediastinum and extends between the first and the last arch vessels (usually between the innominate and the left subclavian arteries).

The descending aorta is the vertical posterior portion of the thoracic aorta, continuing until it reaches the diaphragm. The abdominal aorta extends from the hiatus of the diaphragm until it bifurcates into the common iliac arteries. The main arteries arising from the abdominal aorta are the celiac trunk, the superior and inferior mesenteric arteries, the renal arteries, and the paired lumbar arteries.

2 IMAGING TECHNIQUES

Cardiovascular imaging plays an important role in the diagnosis of Aneurysms-Osteoarthritis syndrome (AOS) and in further monitoring and management of the disease. Imaging tools are used to evaluate the dimensions of the aorta and

Aneurysms-Osteoarthritis Syndrome. http://dx.doi.org/10.1016/B978-0-12-802708-0.00006-5
Copyright © 2017 Elsevier Inc. All rights reserved.

other small and medium-size arteries throughout the arterial tree. Because the aortic diameter threshold in terms of risk of rupture appears to be smaller compared to other connective tissue diseases, monitoring the disease's progression via imaging is very important. Furthermore, cardiac function and morphology are assessed in patients with AOS. To assess these different aspects, several imaging modalities are available, which are discussed in this section (Fig. 5.1). Table 5.1 shows the strengths and weaknesses for the different modalities that can be considered for AOS imaging [1].

2.1 Echocardiography

Transthoracic echocardiography is the first choice for imaging the aorta, because it is safe, widely available, and easy to perform, and it allows for diverse visualization of the aortic root and proximal descending aorta. Aortic diameters should be measured from inner edge to inner edge [2]. However, echocardiography is operator-dependent; some imaging technicians may have difficulty assessing the aortic arch and the descending aorta. These limitations can be overcome by the use of transesophageal echocardiography, which can visualize the aortic arch and the descending aorta. A short segment of the ascending aorta that lies close to the arch cannot be visualized due to the interposition of the right bronchus and the trachea. However, transesophageal echocardiography requires a skilled operator.

AOS is also known to be associated with congenital heart defects in a higher frequency than in the general population, such as a bicuspid aortic valve, atrial septal defect, and persistent ductus arteriosus [3]. Furthermore, left ventricular hypertrophy, mitral valve abnormalities (regurgitation, prolapse), and left atrial enlargement have been reported. [3] Therefore, transthoracic echography is an essential tool to assess cardiac function and morphology at a patient's initial presentation.

2.2 Computed Tomography

Computed tomography may be used for screening and follow-up of the entire aorta and vascular anatomy from head to toe. Drawbacks of computed tomography includes its use of radiation and an intravenous injection of iodine-containing contrast, which can be nephrotoxic. Computed tomography is independent of patient characteristics or patient cooperation and has become the most applied technique for visualization of the arterial tree. Due to technical developments, the radiation and contrast doses given to the patient have decreased sufficiently, but they are still a point of concern, especially in a patient with AOS who requires frequent follow-up scans throughout his or her entire life. To reduce the radiation dose, the acquisition should be made in only a single cardiac phase. It is important that the same cardiac phase be chosen for scanning at each follow-up, as there can be a difference of up to 5 mm in aortic diameter between systole and diastole [4]. The measurements are performed in a sagittal oblique orientation (the so-called candy cane orientation; Fig. 5.1A). Also, the imager must be aware of different

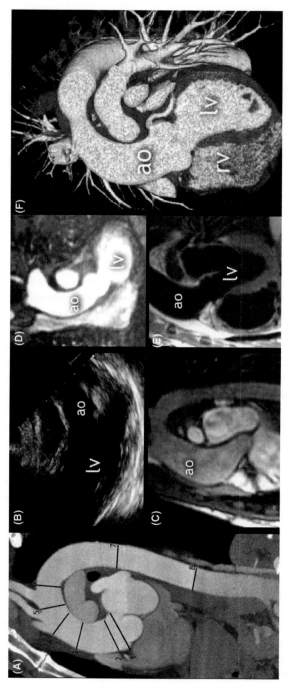

FIGURE 5.1 **Anatomy of thoracic aorta visualized with non-invasive imaging modalities.** Normal aortic anatomy visualized with contrast-enhanced computed tomography (panel B), transthoracic echocardiography (panel B), contrast-enhanced magnetic resonance imaging (panel C), noncontrast magnetic resonance imaging (panels D and E), and three-dimensional reconstruction of contrast-enhanced computed tomography (panel F). Note in panel A the double oblique (candy-cane) reformat image of the aorta and the specific sites for measuring aortic segments: *1*, level of sinus of Valsalva; *2*, sinotubular junction; *3*, ascending aorta at the level of pulmonary artery bifurcation; *4*, ascending aorta before the aortic arch level; *5*, aortic arch; *6*, descending aorta after aortic arch level; *7*, descending thoracic aorta mid-level; *8*, descending aorta at the level of diaphragm. *Ao*, aorta; *lv*, left ventricle; *rv*, right ventricle.

TABLE 5.1 Comparison of methods for imaging of the aorta

Modality	TTE	TEE	CT	MRI
Operator independency	++	+++	++	+++
Diagnostic reliability	+	+++	+++	+++
Bedside use	++	++	−−	−−
Reproducibility	+	+++	+++	+++
Serial examinations	++	+	+++	+++
Radiation dose	0	0	++	0
Contrast	0	0	Obligatory	Optional
Costs	+	++	++	+++

TTE, transthoracic echocardiography; TEE, transesophageal echocardiography; CT, computed tomography; MRI, magnetic resonance imaging.
Source: Adapted from the ESC guidelines on the diagnosis and treatment of aortic diseases [1].

factors that might influence the measurements, such as motion, streak, or metallic artefacts. Furthermore, in patients suspected of having acute aortic syndrome, a computed tomography scan can be used for a fast and reliable diagnosis.

When prophylactic surgical repair of an aortic aneurysm is considered, preprocedural precise and accurate information about the length and width of the diseased segment and its relationship with the side branches is needed. Computed tomography is the most frequently used technique for preprocedural imaging, as it provides accurate and detailed anatomical information and can detect regions with calcifications.

2.3 Magnetic Resonance Imaging

Magnetic resonance imaging (MRI) is based on the magnetic properties of tissues and applies magnetic field and radiofrequency waves for anatomical, functional, and even molecular imaging. In terms of noninvasive imaging, magnetic resonance is the only technique that can image the entire arterial tree without the use of ionizing radiation and the contrast agent. Different sequences visualize the aorta, such as "black blood," "white blood," time of flight, and contrast-enhanced magnetic resonance angiography (Fig. 5.1D and E). For aortic diameter measurements, leading edge to leading edge is commonly used [5].

In addition, MRI can also assess the cardiac morphology. MRI can be performed during the entire cardiac cycle and can quantify shunt fraction in the presence of a ventricular or atrial septal defect. However, MRI is not able to identify calcifications; it is also more time-consuming and requires a dedicated technician and cooperation from the patient. Furthermore, it can be contraindicated in the presence of a pacemaker and is difficult to perform for claustrophobic patients. Recently, the less-demanding, so-called four-dimensional flow technique has been gaining more attention from software developers and from clinicians, because it images the great vessels and the heart during the entire cardiac cycle without the

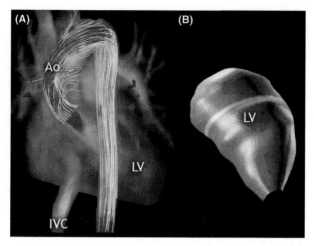

FIGURE 5.2 **Four-dimensional flow magnetic resonance imaging.** Four-dimensional flow magnetic resonance technique provides information about flow (panel A) and heart chambers (panel B) in only one acquisition scan. Panel A, aortic blood flow visualized with streamlines; panel B, volumetric reconstruction of the left ventricle out of the four dimensional flow data for measuring ventricular volume. *Ao*, aorta; *lv*, left ventricle; *ivc*, inferior vena cava.

need for patient cooperation (Fig. 5.2). During a 10-min scan, it provides information about blood flow and heart volumes and function [6].

3 IMAGING OF AORTIC ANEURYSMS

The normal thoracic aortic diameters, which vary according to gender, age, and body surface area, as measured by computed tomography are shown in Table 5.2 [7–9]. Reference values for transthoracic echocardiography corrected for body surface area can be found in the paper by Roman et al [10].

Aortic dilatation is defined as any degree of enlargement, but an aneurysm is defined as dilatation of more than 50% of the normal reference diameter [7]. The risk of rupture increases with the size of the aortic aneurysm [11]. The risk of rupture from aneurysms is greater for women than men [12].

The goal of imaging an aortic aneurysm is to assess its location and maximal diameter [1,7,13]. Over time, the progression rate of dilatation can be measured. Additional characteristics that can be evaluated include morphology (fusiform, saccular), calcifications, filling defects or intramural hematoma, penetrating aortic ulcer, involvement of side branches, signs of malperfusion, and aortic rupture.

Fig. 5.3 shows aortic aneurysms visualized with different imaging techniques. Aneurysm progression is not always linear, and sudden rapid expansion may occur. An annual progression rate of more than 5 mm per year is generally considered indicative of the need for aortic repair or shorter follow-up intervals [3,14]. In AOS patients, the annual progression of the dilated aortic root is highest at the level of the sinus of Valsalva, and dilatation seems to progress at a higher rate compared to that seen with other connective tissue diseases

TABLE 5.2 Ascending and descending aorta reference sizes

Age (years)	BSA(m^2)	Male		Female	
		Ascending	Descending	Ascending	Descending
<45	<1.70	28.6 ± 2.2	20.9 ± NA	28.4 ± 2.7	20.2 ± 1.4
	1.70–1.89	30.1 ± 3.1	22.6 ± 2.0	30.0 ± 2.2	21.4 ± 1.6
	1.90–2.09	30.9 ± 2.7	23.3 ± 1.7	29.8 ± 2.6	20.3 ± 1.2
	>2.1	32.3 ± 3.0	24.3 ± 2.0	31.3 ± NA	21.9 ± NA
45–54	<1.70	31.0 ± 3.8	22.0 ± 1.1	29.6 ± 2.8	21.1 ± 1.6
	1.70–1.89	31.7 ± 3.2	23.5 ± 2.0	31.4 ± 2.9	22.2 ± 1.6
	1.90–2.09	33.1 ± 3.3	24.8 ± 2.2	32.5 ± 3.2	23.6 ± 1.8
	>2.1	34.4 ± 3.1	25.8 ± 1.9	34.4 ± 3.1	23.9 ± 2.2
55–64	<1.70	31.5 ± 2.4	23.1 ± 1.5	31.1 ± 2.9	22.3 ± 1.8
	1.70–1.89	33.5 ± 3.1	25.2 ± 1.7	31.8 ± 2.6	23.3 ± 1.9
	1.90–2.09	34.6 ± 3.3	25.9 ± 2.0	33.0 ± 3.0	24.0 ± 1.9
	>2.1	36.1 ± 3.5	27.2 ± 2.2	35.4 ± 3.3	25.5 ± 3.1
≥65	<1.70	33.9 ± 2.3	25.3 ± NA	32.5 ± 2.5	23.4 ± 1.8
	1.70–1.89	35.0 ± 3.0	26.8 ± 2.8	33.4 ± 2.9	24.6 ± 1.4
	1.90–2.09	35.8 ± 3.2	27.0 ± 2.0	34.3 ± 4.2	25.2 ± 1.9
	>2.1	36.8 ± 2.8	28.5 ± 2.0	32.8 ± NA	26.0 ± 1.9

Normal values for aortic dimensions on noncontrast CT, at pulmonary trunk bifurcation level.
Source: Adapted from Wolak at al. [8].

[3,14]. For more details about clinical and imaging follow-up of AOS patients Chapter 6f.

4 IMAGING OF SMALL AND MEDIUM-SIZE VESSELS

AOS is known to cause aneurysms throughout the entire arterial tree (Fig. 5.4).

4.1 Visceral Arteries

In AOS patients, aneurysms of the visceral arteries have been reported mainly at the level of the celiac, splenic, hepatic, superior mesenteric, and iliac arteries (Fig. 5.4C). [3,15] Computed tomography angiography is usually the imaging technique of choice for follow-up of visceral aneurysms, because it provides detailed images due to its high spatial and temporal resolution. For more details about management and intervention options and thresholds Chapter 6c.

4.2 Cerebral Arteries

Prevalence of intracranial aneurysms and tortuosity is higher in AOS patients when compared with patients with Marfan, Loeys-Dietz, vascular Ehlers-Danlos,

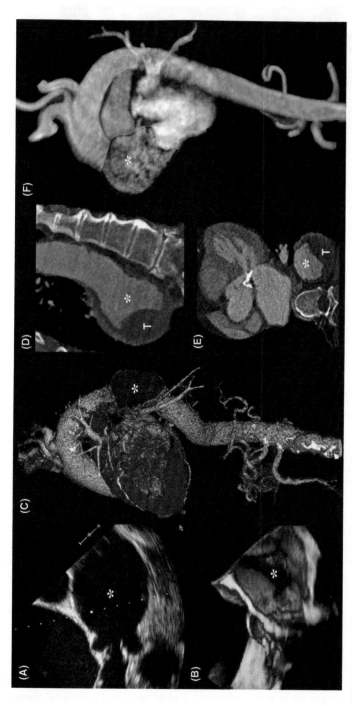

FIGURE 5.3 **Aortic aneurysms displayed with different imaging modalities.** Aneurysm (*asterisk*) of the ascending (panels A, B, and F) and of the descending aorta (panels C, D, and E), visualized with two-dimensional transesophageal echocardiography (panel A), three-dimensional transesophageal echocardiography (panel B), contrast-enhanced computed tomography (panels C, D, and E), and contrast-enhanced magnetic resonance imaging (panel F). Note the presence of thrombus (T) in panels D and E.

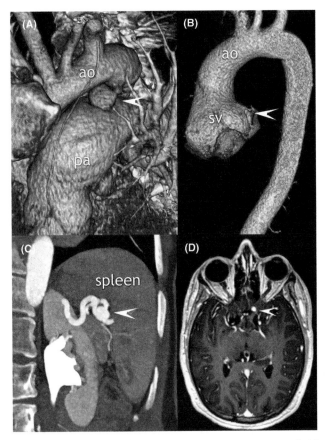

FIGURE 5.4   **Aneurysms throughout the arterial tree in aneurysms-osteoarthritis syndrome.**
Panel A, aneurysm *(arrow head)* of the ductus of Botalli; panel B, aneurysm *(arrow head)* of sinus
of Valsalva; panel C, aneurysm *(arrow head)* of splenic artery; panel D, aneurysm *(arrow head)* of
left internal carotid artery. *Ao*, aorta; *pa*, pulmonary artery; *sv*, sinus of Valsalva.

and Turner syndromes [3]. In a small cohort study, the majority of aneurysms
were located in the basilar artery, while tortuous changes were also found in the
internal carotid, vertebral, cerebral anterior, media, and vertebrobasilar arteries
[3]. Furthermore, aneurysms have been described in the carotid, ophthalmic, and
vertebral arteries (Fig. 5.4D). Depending on their availability, computed tomography and MRI are suitable for follow-up of intracranial and vertebral aneurysms.

5   IMAGING OF AORTIC DISSECTIONS

Aortic dissection represents a tear in the intima layer of the aortic wall and can
be diagnosed with multiple imaging techniques (Fig. 5.5). While blood enters
and accumulates between the vessel wall layers, a cavity is formed, called the
false lumen. Prognosis is worse when the ascending aorta is involved, largely

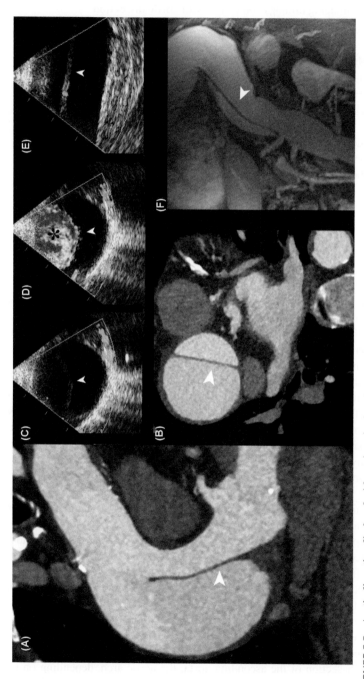

FIGURE 5.5 **Aortic dissection displayed with noninvasive imaging modalities.** Dissection flap (*arrow head*) visualized with contrast-enhanced computed tomography (panels A and B), with transesophageal echocardiography (panels C, D, and E), and with contrast-enhanced magnetic resonance imaging (panel F). Note in panel D the presence of Doppler flux (*asterisk*) in the true lumen and the absence of flux in the false lumen.

TABLE 5.3 Imaging findings associated with aortic dissection

• Break sign	Acute angle between outside wall of the false lumen and intimal flap
• Cobwebs	Filamentous stranding into the false lumen
• Wrapping at the level of the arch	The lumen that wraps around the other is the false one
• Lumen size	False lumen is generally wider except for the most proximal part of the dissection
• Outer wall calcification	Useful to identify the true lumen only in acute dissection; later also the false lumen may calcify
• Thrombosis	Typical for false lumen in about half of the cases; thrombosis may be present within the true lumen in acute dissection
• Direction of the flap free edges	The free edges of the intimal flap on the margins of the primary tear point into the false lumen, passively directed by flow entering the false lumen

due to the increased risk of acute pericardial tamponade. The Stanford classification of aortic dissections is the most frequently used, which differentiates based on the involvement of the ascending aorta [16]. Type A represents all dissections that include the ascending aorta, and type B does not include the ascending aorta. In young patients presenting with aortic dissection, AOS should be considered if no underlying pathology is found.

Computed tomography is the diagnostic tool of choice, when available, because it is fast and highly accurate at detecting the intimal flap [1,7,17]. Usually, the true lumen can be followed from the proximal part of the dissection. Characteristics that help identify an aortic dissection are shown in Table 5.3. Computed tomography scans without the use of an intravascular contrast medium are better for detecting intramural hematoma, mediastinal hemorrhage, and hemorrhagic pericardial or pleural effusion. The measured computed tomography Hounsfield units of hematomas are generally higher than those of the arterial blood pool. Due to motion artefacts that may mimic aortic dissection at the level of the ascending aorta, computed tomography imaging using electrocardiographic synchronization is highly recommended. Other type of artefacts that may mimic aortic dissection may arise from suboptimal timing of contrast administration.

Magnetic resonance angiography is not frequently applied in acute situations, but it is a good option for imaging chronic aortic dissections, particularly in younger patients who need repeated follow-up imaging.

Transthoracic echocardiography is a fast, widely available screening tool. If computed tomography is not available, transthoracic echocardiography can identify a dissection of the ascending aorta; the aortic arch, depending on the suprasternal acoustic window; functional compromise of the aortic valve; and pericardial effusion. Transesophageal echocardiography allows for excellent imaging of the ascending aorta and has good sensitivity and specificity to detect an ascending aortic dissection. However, it can may difficult and perhaps unsafe to perform transesophageal echocardiography on unstable patients [1,7].

A simple chest X-ray usually will not be normal, but it is also not specific for aortic dissection. Findings may include widened superior mediastinum, intimal calcium displacement, or pulmonary venous congestion, which suggests aortic regurgitation, pericardial tamponade, or myocardial infarction. Furthermore, left pleural effusion may be a predictor for rupture.

6   IMAGING AFTER AORTIC SURGERY

After aortic root surgery, imaging in regular intervals is indicated [18]. Interpretation of the postsurgical images has to be done in concordance with the surgical report. Postoperatively, imaging shows the patency of the graft. Aortic tissue grafts are indistinguishable on computed tomography imaging from native aortic tissue. The elephant trunk surgery procedure, by which the graft replaces the arch without an end-to-end anastomosis with the native descending aorta, should not be mistaken for aortic dissection.

Postoperative complications that should be reported include anastomotic dehiscence, contrast leakage (presence of active bleeding), pseudoaneurysms, pericardial and pleural effusion, aortic dissection, infection (mediastinitis, abscess formation), hematoma, aneurysms, or lung injury (pneumothorax) [19]. Furthermore, in patients with AOS, special attention must be paid to the distal (unrepaired) part of the aorta for the development of aneurysms, tortuosity, or dissection [3,18].

REFERENCES

[1] Erbel R, Aboyans V, Boileau C, ESC Committee for Practice Guidelines. et al. 2014 ESC Guidelines on the diagnosis and treatment of aortic diseases: Document covering acute and chronic aortic diseases of the thoracic and abdominal aorta of the adult. The Task Force for the Diagnosis and Treatment of Aortic Diseases of the European Society of Cardiology (ESC). Eur Heart J 2014;35(41):2873–926.

[2] Lang RM, Bierig M, Devereux RB, et al. Recommendations for chamber quantification. Eur J Echocardiogr 2006;7(2):79–108.

[3] van der Linde D, van de Laar IM, Bertoli-Avella AM, et al. Aggressive cardiovascular phenotype of aneurysms-osteoarthritis syndrome caused by pathogenic SMAD3 variants. J Am Coll Cardiol 2012;60(5):397–403.

[4] van Prehn J, Vincken KL, Muhs BE, et al. Toward endografting of the ascending aorta: insight into dynamics using dynamic cine-CTA. J Endovasc Ther 2007;14(4):551–60.

[5] Burman ED, Keegan J, Kilner PJ. Aortic root measurement by cardiovascular magnetic resonance: specification of planes and lines of measurement and corresponding normal values. Circ Cardiovasc Imaging 2008;1(4):104–13.

[6] Vasanawala SS, Hanneman K, Alley MT, Hsiao A. Congenital heart disease assessment with 4D flow MRI. J Magn Reson Imaging 2015;42(4):870–86.

[7] Hiratzka LF, Bakris GL, Beckman JA, et al. 2010 ACCF/AHA/AATS/ACR/ASA/SCA/ SCAI/SIR/STS/SVM guidelines for the diagnosis and management of patients with Thoracic Aortic Disease: a report of the American College of Cardiology Foundation/American Heart Association Task Force on Practice Guidelines, American Association for Thoracic Surgery, American College of Radiology, American Stroke Association, Society of Cardiovascular Anesthesiologists, Society for Cardiovascular Angiography and Interventions, Society of Interventional Radiology, Society of Thoracic Surgeons, and Society for Vascular Medicine. Circulation 2010;121(13):e266–369.

[8] Wolak A, Gransar H, Thomson LE, et al. Aortic size assessment by noncontrast cardiac computed tomography: normal limits by age, gender and body surface area. JACC Cardiovasc Imaging 2008;1(2):200–9.

[9] Hager A, Kaemmerer H, Rapp-Bernhardt U, et al. Diameters of the thoracic aorta throughout life as measured with helical computed tomography. J Thorac Cardiov Surg 2002;123(2): 1060–6.

[10] Roman MJ, Devereux RB, Kramer-Fox R, et al. Two-dimensional echocardiographic aortic root dimensions in normal children and adults. Am J Cardiol 1989;64(8):507–12.

[11] Coady MA, Rizzo JA, Hammond GL, Kopf GS, Elefteriades JA. Surgical intervention criteria for thoracic aortic aneurysms: a study of growth rates and complications. Ann Thorac Surg 1999;67(6):1922–6.

[12] Norman PE, Powell JT. Abdominal aortic aneurysm: the prognosis in women is worse than in men. Circulation 2007;115(22):2865–9.

[13] Isselbacher EM. Thoracic and abdominal aortic aneurysms. Circulation 2005;111(6):816–28.

[14] van der Linde D, Bekkers JA, Mattace-Raso FU, et al. Progression rate and early surgical experience in the new aggressive aneurysms-osteoarthritis syndrome. Ann Thorac Surg 2013;95(2):563–9.

[15] van der Linde D, Verhagen HJ, Moelker A, et al. Aneurysm-osteoarthritis syndrome with visceral and iliac artery aneurysms. J Vasc Surg 2013;57(1):96–102.

[16] Lansman SL, McCullough JN, Nguyen KH, et al. Subtypes of acute aortic dissection. Ann Thorac Surg 1999;67(6):1975–80.

[17] Baliga RR, Nienaber CA, Bossone E, et al. The role of imaging in aortic dissection and related syndromes. JACC Cardiovasc Imag 2014;7(4):406–24.

[18] Prescott-Focht JA, Martinez-Jimenez S, Hurwitz LM, et al. Ascending thoracic aorta: postoperative imaging evaluation. Radiographics 2013;33(1):73–85.

[19] Kari FA, Russe MF, Peter P, et al. Late complications and distal growth rates of Marfan aortas after proximal aortic repair. Eur J Cardiothoracic Surg 2013;44(1):163–71.

Chapter 6

Treatment Options

D. van der Linde, MD, MSc, PhD, J.W. Roos-Hesselink, MD, PhD

1 INTRODUCTION

In the first subchapter, optimal cardiovascular medical treatment strategies for patients with Aneurysms-Osteoarthritis syndrome (AOS) are discussed. For a long time, beta-blockers were the cornerstone treatment aimed at slowing down aneurysm progression in patients with Marfan syndrome and related disorders. However, new insights in the pathogenesis of aortic aneurysm development, such as the role of the transforming growth factor-beta (TGF-β) signaling pathway, have led to new therapeutic targets. Losartan, an angiotensin receptor blocker, has recently received much attention as an alternative treatment option to beta-blockers. Finally, this subchapter briefly discusses promising future treatment options.

The second subchapter discusses the indications, outcome, and recommendations for prophylactic aortic root replacement in patients with AOS. Because the progression of aortic root aneurysms in AOS patients can be fast and unpredictable, with aortic dissections occurring in relatively mildly dilated aortas, early prophylactic surgical intervention should be considered to avoid vascular catastrophes, especially as elective valve sparing aortic root replacement shows favorable results.

AOS is also known to be associated with aneurysms and arterial tortuosity in large and medium-size vessels throughout the arterial tree. Therefore, the third subchapter focuses on endovascular and surgical treatment options for visceral and iliac aneurysms associated with AOS.

The cardiovascular abnormalities in AOS are often asymptomatic for a long time, so most patients initially seek medical attention for joint complaints. AOS is associated with early-onset osteoarthritis, osteochondritis dissecans, and several other musculoskeletal abnormalities. In the fourth subchapter, the musculoskeletal phenotype, clinical evaluation and follow-up, and current orthopedic treatment options are discussed in detail.

The fifth subchapter describes the genetic counseling process, from the facilitation of autonomous decision making to the medical and psychological implications of genetic testing and its possible effects on social and familial

Aneurysms-Osteoarthritis Syndrome. http://dx.doi.org/10.1016/B978-0-12-802708-0.00011-9
Copyright © 2017 Elsevier Inc. All rights reserved.

interactions. The difficult ethical issue of genetic testing in minors is explored. Furthermore, several reproductive options, such as prenatal diagnosis and pre-implantation genetic diagnosis, are discussed.

Finally, the last subchapter provides an approach to patient-centered clinical management for AOS patients in a multidisciplinary team. An algorithm for screening and clinical follow-up recommendations based on all the available knowledge thus far is proposed.

Chapter 6a

Optimal Cardiovascular Medical Treatment

B.L. Loeys, MD, PhD

1 BETA-BLOCKERS: STANDARD OF CARE?

The current optimal medical treatment strategy for patients with Aneurysms-Osteoarthritis syndrome (AOS) is largely based on experience gained from treating Marfan syndrome (MFS) patients. Although controversial, until recently the standard medical treatment for MFS relied upon beta-blockers as the first-line prophylaxis [1]. The therapeutic effect of beta-blockers is obtained through a decrease in blood pressure and heart rate, leading to reduced stress on the aortic wall and, as such, a slowdown of aortic dilatation. In addition, beta-blockers are hypothesized to also induce cross-linking between extracellular matrix components and consequently improve elastic aortic properties [2].

Yet the trial evidence for the effectiveness of beta-blockers in the treatment of MFS patients is rather limited and based only on a single, small, randomized study. Shores et al reported in 1994 on a cohort of 70 MFS patients followed for 9.3 years in the placebo group ($n = 38$) and 10.7 years in the beta-blocker-(propranolol-) treated MFS group. Although the rate of aortic root dilatation was slower in the propranolol-treated group, none of the primary endpoints, including aortic dissection, cardiovascular surgery, or death, were significantly different [3]. In an attempt to provide more solid evidence for the effectiveness of beta-blockers, a metaanalysis was carried out on six published studies [4]. However, due to significant variability in the trial designs (eg, nonrandomized, small size, different beta-blockers), no meaningful conclusions could be drawn.

2 FROM SYMPTOMATIC TO CAUSAL TREATMENT

Over the last decade, several lines of evidence have pointed toward the key role of transforming growth factor-beta (TGF-β) signaling in the development of aortic aneurysm.

First of all, after the initial demonstration of the essential role of TGF-β signaling in the development of lung emphysema in a MFS mouse model [5],

Aneurysms-Osteoarthritis Syndrome. http://dx.doi.org/10.1016/B978-0-12-802708-0.00012-0
Copyright © 2017 Elsevier Inc. All rights reserved.

this signaling was also demonstrated to be driving the aortic aneurysm formation in MFS mice. The use of TGF-β neutralizing antibodies in MFS mice attenuated the aortic root growth [6].

Second, additional evidence for the crucial involvement of TGF-β signaling in the pathogenesis of aortic aneurysm and dissection was derived from the discovery of the genetic basis of a new aortic aneurysm syndrome, Loeys–Dietz syndrome (LDS). It was demonstrated that this syndrome was caused by mutations in the receptors for TGF-β, transforming growth factor receptors 1 and 2 [7], but recently mutations in the genes coding for the ligand TGFB2 [8] and TGFB3 were also associated with LDS-like presentations [9]. Despite the loss-of-function nature of mutations in these genes, studying the cell lines and aortic tissues of LDS patients and mouse models demonstrated a paradoxical increase in TGF-β signaling. Similar observations have been made in the tissues of patients with AOS related to *SMAD3* mutations [10].

All these observations have led to the initiation of a new treatment strategy for aortic aneurysm with angiotensin receptor blockers, especially losartan. In addition to their antihypertensive effect, angiotensin receptor blockers can also attenuate TGF-β signaling by lowering the expression of TGF-β ligands, receptors, and activators [11]. As such, it is considered to be an excellent alternative for the use of neutralizing antibodies. Indeed, losartan has been demonstrated to be capable of slowing the aortic aneurysm development in the MFS mouse models, similar to what was previously observed after the use of neutralizing antibodies [6].

3   ANGIOTENSIN RECEPTOR BLOCKERS: AN ALTERNATIVE TREATMENT STRATEGY?

The recent advances in our understanding of the underlying pathways in aortic aneurysm in MFS and related disorders, along with the very promising data from the losartan treatment in MFS mice, have sparked a lot of enthusiasm and initiated several clinical trials for MFS patients. A first, small, retrospective study in severely affected young MFS patients (ages 1–18 years) whose aortas continued to grow despite optimal beta-blocker treatment (atenolol or propranolol) showed that the addition of angiotensin receptor blocker therapy (losartan, or irbesartan in one patient) significantly reduced the aortic growth at the sinuses of Valsalva with a follow-up between 12 and 47 months [12]. Initially, some smaller trials have confirmed the positive results from this first retrospective study [13,14]. A large, randomized, controlled trial from the Netherlands also demonstrated the beneficial effects of losartan in an adult MFS population [15]. Two other large studies showed equal results for beta-blocker versus angiotensin receptor blocker treatments. In a large trial sponsored by the Pediatric Heart Network ($n = 604$), a classic dose of losartan (average 1.2 mg/kg/day) was equally effective to high dose of atenolol (average 2.7 mg/kg/day) [16]. Remarkably, this high dose of beta-blocker was well tolerated by the trial patients, an observation not routinely seen in daily clinical practice. Another interesting lesson from this trial is the fact that earlier initiation of beta-blockers or losartan leads to slower aortic growth rate compared to treatment started at a later age. Finally, in a French trial, patients using either placebo or losartan on top of their routine treatment (86% took beta-blockers) showed no difference in their aortic growth rate [17].

Although clinicians now seem to have a choice between administering beta-blockers or losartan, several questions remain unanswered. Is a high dose of losartan better than a low, normal dose of losartan? Are new angiotensin receptor blockers, such as irbesartan, more effective than losartan? Until now, all angiotensin receptor blocker studies have been carried out on patients with MFS; no trials for patients with mutations in the TGF-β signaling genes have been performed. Several lines of evidence suggest that angiotensin receptor blocker treatment may also be effective in these patients. Investigation of aortic tissues of these patients showed a paradoxical increase in TGF-β signaling. Angiotensin receptor blocker treatment of LDS mouse models (mutations in tgfbr1/2) showed beneficial effects of losartan on the aortic growth rate, [18] and preliminary results in LDS patients seem to confirm the inhibitory effect of angiotensin receptor blockers on aortic diameter increases [19].

4   FUTURE TREATMENT OPTIONS

A small, randomized trial of angiotensin-converting-enzyme- (ACE-) inhibitor therapy reported a reduction in aortic root and ascending aortic growth in MFS patients [20]. In the fibrillin-1 deficient mouse model, angiotensin receptor blocker therapy (with losartan) was superior to ACE-inhibitor therapy (with enalapril) in preventing aortic enlargement and improving aortic architecture [21], but there are no comparative data in humans. Most recently, a detrimental effect of calcium channel blockers was suggested [22].

One of the emerging treatment targets is the noncanonical TGF-β signaling pathway through extracellular signal-regulated kinase (ERK) signaling. Preliminary mouse experiments have shown that the administration of the potent ERK inhibitor RDEA119 arrests aortic growth in MFS mice [23]. In view of the importance and beneficial effect of angiotensin type 2 receptor (AT2R) mediated signaling, AT2R agonists are attractive treatment options. In addition, alternative pathways, such as the MAS receptor–related pathways (which may have the same beneficial effect as AT2R mediated signaling), offer excellent therapeutic targets.

Other treatment strategies have investigated the role of macrophages and matrix metalloproteinases in MFS mouse models. Both the use of antibodies-blocking GxxPG fragments and doxycycline have demonstrated beneficial effects in MFS mouse models [24–26].

So far, human evidence for the efficacy of doxycycline is limited to the treatment of abdominal aortic aneurysm [27,28].

Finally, some evidence has suggested that statins can ameliorate aortic growth in the MFS mouse model. In a short-term (6 weeks) comparative study, pravastatin and losartan performed equally well [29]. Alternatively, the use of statins in combination with peroxisome proliferators-activated g-receptor (PPARg) agonists, a potent antimatrix metalloproteinase and TGF-β compound, has also been suggested as a possible novel MFS treatment strategy. [30].

REFERENCES

[1] Braverman AC. Timing of aortic surgery in the Marfan syndrome. Curr Opin Cardiol 2004;19(6):549–50.

[2] Brophy CM, Tilson JE, Tilson MD. Propranolol stimulates the crosslinking of matrix components in skin from the aneurysm-prone blotchy mouse. J Surg Res 1989;46(4):330–2.

[3] Shores J, Berger KR, Murphy EA, et al. Progression of aortic dilatation and the benefit of long-term beta-adrenergic blockade in Marfan's syndrome. New Engl J Med 1994;330(19): 1335–41.

[4] Gersony DR, McClaughlin MA, Jin Z, et al. The effect of beta-blocker therapy on clinical outcome in patients with Marfan's syndrome: a meta-analysis. Int J Cardiol 2007;114(3):303–8.

[5] Neptune ER, Frischmeyer PA, Arking DE, et al. Dysregulation of TGF-beta activation contributes to pathogenesis in Marfan syndrome. Nat Genet 2003;33(3):407–11.

[6] Habashi JP, Judge DP, Holm TM, et al. Losartan, an AT1 antagonist, prevents aortic aneurysm in a mouse model of Marfan syndrome. Science 2006;312(5770):117–21.

[7] Loeys BL, Chen J, Neptune ER, et al. A syndrome of altered cardiovascular, craniofacial, neurocognitive and skeletal development caused by mutations in TGFBR1 or TGFBR2. Nat Genet 2005;37(3):275–81.

[8] Lindsay ME, Schepers D, Bolar NA, et al. Loss-of-function mutations in TGFB2 cause a syndromic presentation of thoracic aortic aneurysm. Nat Genet 2012;44(8):922–7.

[9] Bertoli-Avella AM, Gillis E, Morisaki H, et al. Mutations in a TGF-beta ligand, TGFB3, cause syndromic aortic aneurysms and dissections. J Am Coll Cardiol 2015;65(13):1324–36.

[10] van de Laar IM, Oldenburg RA, Pals G, et al. Mutations in SMAD3 cause a syndromic form of aortic aneurysms and dissections with early-onset osteoarthritis. Nat Genet 2011;43(2):121–6.

[11] Fukuda N, Hu WY, Kubo A, et al. Angiotensin II upregulates transforming growth factor-beta type I receptor on rat vascular smooth muscle cells. Am J Hypertens 2000;13(2):191–8.

[12] Brooke BS, Habashi JP, Judge DP, et al. Angiotensin II blockade and aortic-root dilation in Marfan's syndrome. New Engl J Med 2008;358(26):2787–95.

[13] Chiu HH, Wu MH, Wang JK, Lu CW, Chiu SN, Chen CA, et al. Losartan added to beta-blockade therapy for aortic root dilation in Marfan syndrome: a randomized, open-label pilot study. Mayo Clin Proc 2013;88(3):271–6.

[14] Pees C, Laccone F, Hagl M, et al. Usefulness of losartan on the size of the ascending aorta in an unselected cohort of children, adolescents, and young adults with Marfan syndrome. Am J Cardiol 2013;112(9):1477–83.

[15] Groenink M, den Hartog AW, Franken R, et al. Losartan reduces aortic dilatation rate in adults with Marfan syndrome: a randomized controlled trial. Eur Heart J 2013;34(45):3491–500.

[16] Lacro RV, Dietz HC, Sleeper LA, et al. Atenolol versus losartan in children and young adults with Marfan's syndrome. N Engl J Med 2014;371(22):2061–71.

[17] Milleron O, Arnoult F, Ropers J, et al. Marfan Sartan: a randomized, double-blind, placebo-controlled trial. Eur Heart J 2015;36(32):2160–6.

[18] Gallo EM, Loch DC, Habashi JP, et al. Angiotensin II-dependent TGF-beta signaling contributes to Loeys-Dietz syndrome vascular pathogenesis. J Clin Invest 2014;124(1):448–60.

[19] Maccarrick G, Black JH 3rd, Bowdin S, et al. Loeys-Dietz syndrome: a primer for diagnosis and management. Genet Med 2014;16(8):576–87.

[20] Ahimastos AA, Aggarwal A, D'Orsa KM, et al. Effect of perindopril on large artery stiffness and aortic root diameter in patients with Marfan syndrome: a randomized controlled trial. JAMA-J Am Med Assoc 2007;298(13):1539–47.

[21] Habashi JP, Doyle JJ, Holm TM, et al. Angiotensin II type 2 receptor signaling attenuates aortic aneurysm in mice through ERK antagonism. Science 2011;332(6027):361–5.

[22] Doyle JJ, Doyle AJ, Wilson NK, et al. A deleterious gene-by-environment interaction imposed by calcium channel blockers in Marfan syndrome. eLife 2015;4. pii: e08648.

[23] Holm TM, Habashi JP, Doyle JJ, et al. Noncanonical TGFβ signaling contributes to aortic aneurysm progression in Marfan syndrome mice. Science 2011;332(6027):358–61.

[24] Guo G, Muñoz-Garcia B, Ott CE, et al. Antagonism of GxxPG fragments ameliorates manifestations of aortic disease in Marfan syndrome mice. Hum Mol Genet 2013;22(3):433–43.

[25] Chung AW, Yang HH, Radomski MW, et al. Long-term doxycycline is more effective than atenolol to prevent thoracic aortic aneurysm in Marfan syndrome through the inhibition of matrix metalloproteinase-2 and -9. Circ Res 2008;102(8):e73–85.

[26] Xiong W, Knispel RA, Dietz HC, et al. Doxycycline delays aneurysm rupture in a mouse model of Marfan syndrome. J Vasc Surg 2008;47(1):166–72. discussion 172.

[27] Curci JA, Mao D, Bohner DG, et al. Preoperative treatment with doxycycline reduces aortic wall expression and activation of matrix metalloproteinases in patients with abdominal aortic aneurysms. J Vasc Surg 2000;31(2):325–42.

[28] Mosorin M, Juvonen J, Biancari F, et al. Use of doxycycline to decrease the growth rate of abdominal aortic aneurysms: a randomized, double-blind, placebo-controlled pilot study. J Vasc Surg 2001;34(4):606–10.

[29] McLoughlin D, McGuinness J, Byrne J, et al. Pravastatin reduces Marfan aortic dilation. Circulation 2011;124(Suppl. 11):S168–73.

[30] Sorice GP, Folli F. A combination of PPAR-gamma agonists and HMG CoA reductase inhibitors (statins) as a new therapy for the conservative treatment of AAS (aortic aneurysm syndromes). Med Hypotheses 2009;73(4):614–8.

Chapter 6b

Cardiothoracic Surgical Experience

D. van der Linde, MD, MSc, PhD, J.A. Bekkers, MD, PhD

1 INTRODUCTION

Aortic aneurysms and dissections are common conditions, ranking as the 19th-most-common causes of death in the United States [1]. The predilection for thoracic aortic aneurysms and dissections can be inherited in an autosomal dominant manner, of which the recently described Aneurysms-Osteoarthritis syndrome (AOS) is an example [2–7]. This syndrome has been found to be responsible for approximately 2% of familial thoracic aortic aneurysms and dissections [3,7].

Aneurysms associated with AOS most commonly occur at the level of the sinuses of Valsalva but can be present throughout the entire arterial tree [3–7]. Aortic dissections can occur in relatively mildly dilated aortas and are associated with a high mortality rate [3–5]. In the ideal situation, one would like to intervene with low-risk surgery before dissections occur. However, in other disorders affecting the aorta—for example, vascular-type Ehlers-Danlos syndrome (EDS)—fragility of the aortic tissue may complicate surgical intervention [8]. When considering prophylactic surgery to prevent aneurysms from rupturing, it is important to know whether friable vascular tissue is also present in AOS patients. This subchapter summarizes the cardiothoracic surgical experience thus far in AOS patients.

2 INDICATIONS FOR ELECTIVE AORTIC ROOT SURGERY

The goal of elective thoracic aortic aneurysm surgery is to prevent the most feared complication of asymptomatic aortic aneurysms: aortic dissection or rupture. This concept might come across as being very simple and logical, but estimating the actual risk of an aortic dissection in each individual patient is very complicated. Limited data are available to make an evidence-based prediction of the risk of aortic dissection. Thus, the decision to undergo aortic surgery is typically based on a combination of predictors, including absolute aortic aneurysm diameter, rate of progression, valve function, and family history.

Aneurysms-Osteoarthritis Syndrome. http://dx.doi.org/10.1016/B978-0-12-802708-0.00013-2
Copyright © 2017 Elsevier Inc. All rights reserved.

123

First of all, it is important to become aware of the data on thoracic aortic dissections in patients with AOS. Although they are extensively described in Chapter 2, we will quickly recapitulate these data now. Several studies have found that if natural history takes its course, about one in three patients will suffer an aortic dissection in the fourth decade of life [3–5,7,9]. Around three-quarters of these dissections have been Stanford type A [3–5]. The range of aortic root measurements before type A aortic dissection occurred was 40–63 mm, with a mean of 51 mm [4]. Occasionally, aortic dissection has occurred in aortas that were only mildly dilated, with evidence of two patients dissecting at diameters of 40 and 45 mm, respectively [4]. Aortic dissections were found to be cause of death in six patients with AOS who died suddenly between the ages of 34 and 69 years [4]. In addition, it appears that some families are more prone to dissections than other families, although the phenotype–genotype relationship has not yet been unraveled [5].

As previously mentioned in Chapter 2, the annual rate of progression of thoracic aortic aneurysm dilatation in AOS has been found to be highest in the sinuses of Valsalva, at approximately 2.5 mm per year [6]. Although this estimate is based on a limited number of patients and should be confirmed in the future by larger studies with longer follow-up intervals, it has become clear that aortic growth in patients with AOS can be fast and unpredictable. The annual progression rate seems comparable to or even higher than in Marfan syndrome (MFS) patients, with progression ranging from 0.4 to 2.1 mm per year [10–14]. Similar to MFS patients, the baseline aortic diameter has also been correlated to progressive aortic dilatation in AOS patients [15]. In patients with a bicuspid aortic valve, the progression of ascending aortic dilatation seems to be lower, with a large variety ranging from 0.2 to 1.9 mm per year [16–19]. Unfortunately, longitudinal studies involving Loeys-Dietz syndrome (LDS) or vascular-type EDS patients are not available, so it is not possible to compare the progression rate results with those of patients with these syndromes.

3 SURGICAL TECHNIQUE FOR VALVE-SPARING AORTIC ROOT REPLACEMENT

Several techniques for valve-sparing aortic root replacement have been described. In our department, replacement of the entire aortic root with reimplantation of the aortic valve, originally described by David et al., and modified by De Paulis et al., is the preferred surgical method [20,21]. In short, after transection of the ascending aorta, the coronary ostia are mobilized and the sinuses of Valsalva are excised, leaving a 2-mm rim of aortic wall above the attachment of the aortic valve leaflets. A specifically designed tubular Dacron graft with pseudo-sinuses of Valsalva is proximally secured with sutures, placed in a horizontal plane in the left ventricular outflow tract, underneath the attachment of the

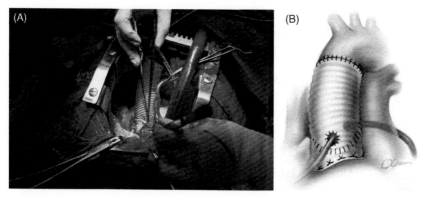

FIGURE 6B.1 **Valve sparing aortic root replacement.** (A) Picture taken in the operating room while the coronaries are reimplanted. (B) Schematic drawing demonstrating the result after the coronary reimplantation. *(Reused with permission from Matalanis G. Valve sparing aortic root repairs—an anatomical approach. Heart Lung Circ 2004;13:S13–S18.)*

aortic valve leaflets. The detached aortic valve is reimplanted inside the prosthesis with continuous sutures. After reimplantation of the coronary buttons, a distal anastomosis between the prosthesis and the ascending aorta is constructed (Fig. 6b.1).

4 ELECTIVE VALVE-SPARING AORTIC ROOT REPLACEMENT EXPERIENCE

The Dutch cohort of AOS patients is the only cohort who specifically described the outcomes of elective aortic root surgeries in detail [6]. Ten patients underwent elective valve-sparing aortic root replacement (age 38.4 ± 14.7 years; 60% male). Preoperative aortic root diameter at the level of the sinuses of Valsalva was 46.6 ± 4.0 mm. Mean cardiopulmonary bypass and cross-clamp times were 168 ± 12 and 141 ± 17 min, respectively. During surgery, aortic fragments were obtained. Histopathology examination showed characteristic loss and fragmentation of elastic fibers, with mucoid medial degeneration in 70% of the fragments.

The operation was uncomplicated in all patients, and all aortic valves could be saved. One patient developed a complete atrioventricular block postoperatively, requiring a permanent pacemaker implantation. In none of the patients were rethoracotomies necessary. One patient had two episodes of paroxysmal atrial flutter 3 weeks postoperatively that was treated successfully with beta blockade. No patients experienced postoperative infections, thromboembolism, or endocarditis. After a postoperative period of 3 years, no mortality or reoperations had occurred, no significant aortic regurgitation had developed, and all patients remained in New York Heart Association functional class I.

These favorable preliminary results seem comparable to the excellent results of valve-sparing aortic root replacement in other patients with aortic aneurysms, including MFS and LDS patients [22–28], but they need to be confirmed in larger series. Unlike patients with vascular-type EDS, who have a high incidence of intraoperative and early postoperative vascular events due to their fragile arterial tissue [8], AOS patients did not have these complications, nor did their aortic tissue feel extremely thin or fragile [6].

5   RECOMMENDATIONS FOR ELECTIVE AORTIC ROOT SURGERY

Although data are scarce thus far and more longitudinal, larger follow-up studies are necessary, at this point it is reasonable to conclude that AOS patients are at high risk for acute aortic dissection with an associated high mortality rate. Furthermore, aortic growth can be fast and unpredictable. Because aortic dissections have been reported to occur in relatively mildly dilated aortas, and elective valve-sparing aortic root replacement shows favorable results, early prophylactic surgical intervention should be considered to avoid vascular catastrophes [6]. As AOS highly resembles LDS with regard to aortic aneurysms and dissections, we suggest applying the current surgical recommendations for LDS: aortic root diameter > 42 mm or rapidly expanding (> 0.5 cm over 1 year) [6,29]. Individualized assessment of risk versus benefit, based on family history and patient characteristics, should always be taken into account.

For postoperative surveillance, we recommend transthoracic echocardiography at 6 months postoperatively and annually thereafter to monitor aortic root diameter and valve competence [6]. Given the widespread involvement of the arterial tree in AOS patients, repeated head-to-pelvis imaging in patients after valve-sparing aortic root replacement remains crucial to evaluate other large and medium-size arteries [3–6].

6   CONCLUSIONS

In view of the aggressive and progressive nature of the thoracic aortic aneurysms associated with AOS and the low rate of complications associated with valve-sparing aortic root replacement surgery at experienced centers, early prophylactic surgery is recommended to avoid catastrophic aortic dissections. Consensus is to recommend surgery if the aortic root diameter exceeds 42 mm or in the case of rapid expansion > 0.5 cm per year. However, in each individual case, the decision to undergo aortic surgery should be made carefully, weighing the individual's aortic aneurysm size and progression rate, valve function, family history, and possible risks.

REFERENCES

[1] National Center for Injury Prevention and Control. WISQARS Leading Causes of Death Reports 2007. Available from: http://webappa.cdc.gov/sasweb/ncipc/leadcaus10.html.

[2] Milewicz DM, Chen H, Park ES, et al. Reduced penetrance and variable expressivity of familial thoracic aortic aneurysms/dissections. Am J Cardiol 1998;82(4):474–9.

[3] van de Laar IM, Oldenburg RA, Pals G, et al. Mutations in SMAD3 cause a syndromic form of aortic aneurysms and dissections with early-onset osteoarthritis. Nat Genet 2011;43(2):121–6.

[4] van der Linde D, van de Laar IM, Bertoli-Avella AM, et al. Aggressive cardiovascular phenotype of aneurysms-osteoarthritis syndrome caused by pathogenic SMAD3 variants. J Am Coll Cardiol 2012;60(5):397–403.

[5] van de Laar IM, van der Linde D, Oei EH, et al. Phenotypic spectrum of the SMAD3-related aneurysms-osteoarthritis syndrome. J Med Genet 2012;49(1):47–57.

[6] van der Linde D, Bekkers JA, Mattace-Raso FU, et al. Progression rate and early surgical experience in the new aggressive aneurysms-osteoarthritis syndrome. Ann Thorac Surg 2013;95(2):563–9.

[7] Regalado ES, Guo DC, Villamizar C, et al. Exome sequencing identifies SMAD3 mutations as a cause of familial thoracic aortic aneurysm and dissection with intracranial and other arterial aneurysms. Circ Res 2011;109(6):680–6.

[8] Oderich GS, Pannenton JM, Bower TC, et al. The spectrum, management and clinical outcome of Ehlers-Danlos syndrome type IV: a 30-year experience. J Vasc Surg 2005;42(6):98–106.

[9] Aubart M, Gobert D, Aubart-Cohen F, et al. Early-onset osteoarthritis, Charcot-Marie-Tooth like neuropathy, autoimmune features, multiple arterial aneurysms and dissections: an unrecognized and life threatening condition. PLoS One 2014;9(5):e96387.

[10] Salim MA, Alpert BS, Ward JC, Pyeritz RE. Effect of beta-adrenergic blockade on aortic root rate of dilatation in the Marfan's syndrome. Am J Cardiol 1994;74(6):629–33.

[11] Kornbluth M, Schnittger I, Eyngorina I, Gasner C, Liang DH. Clinical outcome in the Marfan syndrome with ascending aortic dilatation followed annually by echocardiography. Am J Cardiol 1999;84(6):753–5.

[12] Meijboom LJ, Timmermans J, Zwinderman AH, Engelfriet PM, Mulder BJ. Aortic root growth in men and women with the Marfan's syndrome. Am J Cardiol 2005;96(10):1441–4.

[13] Lazarevic AM, Nakatani S, Okita Y, et al. Determinants of rapid progression of aortic root dilatation and complications in Marfan syndrome. Int J Cardiol 2006;106(2):177–82.

[14] Jondeau G, Detaint D, Tubach F, et al. Aortic event rate in the Marfan population: a cohort study. Circulation 2012;125(2):226–32.

[15] Nollen GJ, Groenink M, Tijssen JG, van der Wall EE, Mulder BJ. Aortic stiffness and diameter predict progressive aortic dilatation in patients with Marfan syndrome. Eur Heart J 2004;25(3):1146–52.

[16] Ferencik M, Pape LA. Changes in size of ascending aorta and aortic valve function with time in patients with congenitally bicuspid aortic valves. Am J Cardiol 2003;92(1):43–6.

[17] Davies RR, Kaple RK, Mandapati D, et al. Natural history of ascending aortic aneurysms in the setting of an unreplaced bicuspid aortic valve. Ann Thorac Surg 2007;83(4):1338–44.

[18] Thanassoulis G, Yip JW, Filion K, et al. Retrospective study to identify predictors of the presence and rapid progression of aortic dilatation in patients with bicuspid aortic valves. Nat Clin Pract Card 2008;5(12):821–8.

[19] van der Linde D, Yap SC, van Dijk AP, et al. Effects of rosuvastatin on progression of stenosis in adult patients with congenital aortic stenosis (PROCAS trial). Am J Cardiol 2011;108(2):265–71.

[20] David TE, Ivanov J, Armstrong S, Feindel CM, Webb GD. Aortic valve-sparing operations in patients with aneurysms of the aortic root and ascending aorta. Ann Thorac Surg 2002;74(5):S1758–61.

[21] David TE, Feindel CM. An aortic valve-sparing operation for patients with aortic incompetence and aneurysm of the ascending aorta. J Thorac Cardiov Sur 1992;103(4):617–21.

[22] de Paulis R, Scaffa R, Nardella S, et al. Use of the Valsalva graft and long-term follow-up. J Thorac Cardiov Sur 2010;140(Suppl. 6):S23–7.

[23] David TE, Feindel CM, Webb GD, Colman JM, Armstrong S, Maganti M. Long-term results of aortic valve-sparing operations for aortic root aneurysm. J Thorac Cardiov Sur 2006;132(2):347–54.

[24] Williams JA, Loeys BL, Nwakanma LU, et al. Early surgical experience with Loeys-Dietz: a new syndrome of aggressive thoracic aortic aneurysm disease. Ann Thorac Surg 2007;83(2):S757–63.

[25] Settepani F, Szeto WY, Pacini D, et al. Reimplantation valve-sparing aortic root replacement in Marfan syndrome using the Valsalva conduit: an intercontinental multicenter study. Ann Thorac Surg 2007;83(2):S769–73.

[26] Kallenbach K, Baraki H, Khaladj N, et al. Aortic valve-sparing operation in Marfan syndrome: what do we know after a decade? Ann Thorac Surg 2007;83(2):S764–8.

[27] Patel ND, Arnaoutakis GJ, George TJ, et al. Valve-sparing aortic root replacement in Loeys-Dietz syndrome. Ann Thorac Surg 2011;92(2):556–60.

[28] Arabkhani B, Mookhoek A, di Centa I, et al. Reported outcome after valve sparing aortic root replacement for aortic root aneurysm: a systematic review and meta-analysis. Ann Thorac Surg 2015;100(3):1126–31.

[29] Hiratzka LF, Bakris GL, Beckman JA, et al. ACCF/AHA/AATS/ACR/ASA/SCA/SCAI/SIR/STS/SVM guidelines for the diagnosis and management of patients with thoracic aortic disease. J Am Coll Cardiol 2010;55(14):e27–e129.

Chapter 6c

Vascular Interventions
and Surgical Experience

D. van der Linde, MD, MSc, PhD, H.J.M. Verhagen, MD, PhD

1 INTRODUCTION

Visceral and iliac aneurysms are relatively rare yet potentially catastrophic when rupturing [1–5]. Although most visceral and iliac aneurysms are degenerative, they can also be encountered in the setting of connective tissue disorders, such as Loeys-Dietz syndrome (LDS) and vascular-type Ehlers-Danlos syndrome (EDS) [5–9]. Arterial aneurysms in patients with Aneurysms-Osteoarthritis syndrome (AOS) are also not limited to the aorta only but may be present throughout the arterial tree in large and medium-size vessels [9–11]. On computed tomography (CT) or magnetic resonance imaging, one-third of patients were found to have aneurysms in other thoracic and abdominal arteries, predominantly involving the pulmonary, splenic, iliac, and mesenteric arteries [5,9–11].

Vascular specialists, such as vascular surgeons and interventional radiologists, should bear in mind the possibility of AOS as a potential underlying cause of visceral and iliac aneurysms or tortuosity, especially in patients with aortic aneurysms or dissections, joint complaints, multiple arterial aneurysms, or a strong family history of aortic dissections or sudden death.

2 INVOLVEMENT OF THE VISCERAL AND ILIAC ARTERIES IN ANEURYSMS-OSTEOARTHRITIS SYNDROME

As described in Chapter 2, one-third of AOS patients have been found to have aneurysms in medium-size arteries throughout the arterial tree [5,9–11]. Fig. 6c.1 shows the distribution and frequency of a total of 73 aneurysms within 17 patients with AOS in a Dutch cohort [5]. Although aneurysms were encountered in a variety of arteries, iliac and splenic artery aneurysms were usually the largest and therefore most frequently required treatment [5]. In addition, arterial tortuosity was also most common in the splenic and iliac arteries [5].

Aneurysms-Osteoarthritis Syndrome. http://dx.doi.org/10.1016/B978-0-12-802708-0.00014-4
Copyright © 2017 Elsevier Inc. All rights reserved.

129

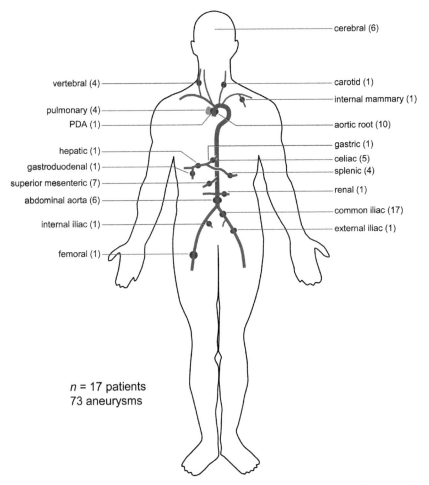

FIGURE 6C.1   **Distribution of 73 aneurysms within 17 patients with Aneurysms-Osteoarthritis syndrome.** *(PDA,* Patent ductus arteriosus) *(Reused with permission from the article by van der Linde et al. Aneurysm-osteoarthritis syndrome with visceral and iliac artery aneurysms. J Vasc Surg 2013;57:96–102.)*

The Dutch cohort is not the only cohort to describe aneurysms in medium-size arteries in patients with AOS; a French cohort of 50 AOS patients reported that 39% had aneurysms in the visceral arteries (renal, splenic, and iliac) or subclavian arteries [12]. It also reported a tortuosity rate of 11% in the visceral and iliac arteries [12].

Although widespread arterial tortuosity and aneurysms can also be found in patients with LDS or arterial tortuosity syndrome, they are rare in patients with Marfan syndrome (MFS) [5,13,14]. Therefore, this characteristic can be helpful in distinguishing AOS patients from MFS patients, although not with 100% sensitivity.

3 VISCERAL AND ILIAC ARTERY ANEURYSM GROWTH

As aneurysm growth can be fast and unpredictable in AOS patients [15], it is unknown how often CT or magnetic resonance angiography (MRA) scans should be repeated.

Repeated CT or MRA scans were only available for six patients in the Dutch cohort, and rapid growth of an aneurysm was noted in two patients [5]. A 30-year-old man had a 6-mm fusiform aneurysm in the left proximal vertebral artery, which increased to 11 mm within 10 months. An aneurysm in the right hepatic artery in a 43-year-old man increased in size from 11 to 18 mm in 9 months' time. Furthermore, a completely new fusiform aneurysm (15 × 11 mm) in the left gastric artery developed within 11 months' time. In four patients, no aneurysmal growth was found in 1–3 years.

Thus, data on visceral and iliac aneurysm growth are scarce for AOS patients so far [5,9–11]. Therefore, the frequency of imaging studies should be based on the individual patient. Until more data are available, we advise repeating the imaging every 1–3 years [5].

4 OPEN VASCULAR INTERVENTIONS

The Dutch cohort of AOS two patients underwent open vascular interventions for peripheral vascular interventions [5,16].

The first patient, a 32-year-old man with bilateral large iliac artery aneurysms (69 mm on the right side and 42 mm on the left side), underwent aortobiiliac graft implantation (Gelsoft prosthesis; Vascutek, Renfrewshire, UK) [5,16]. The quality of the tissue seemed normal, and no perioperative complications occurred. A follow-up MRA scan at 1 month showed a relative stenosis of the distal anastomosis due to progressive tortuosity and elongation of the native common and external iliac arteries. Fig. 6c.2 shows the MRA scans before and after surgery [5,16]. Three months later, the patient required a mesh repair of an incisional hernia, which was likely related to abnormal collagen composition due to AOS.

The second patient, a 35-year-old man, underwent surgical resection with end-to-end anastomosis of a splenic artery aneurysm with a diameter of 60 mm [5]. No perioperative complications occurred.

5 ENDOVASCULAR INTERVENTIONS

Three patients in the Dutch AOS cohort underwent endovascular interventions [5]. Two patients with splenic artery aneurysms (diameter 20–25 mm) underwent coil embolization with occlusion of the splenic artery (Fig. 6c.3) [5]. Both patients complained of postprocedural abdominal pain for some days, most likely due to splenic ischemia, which was adequately managed with analgesics.

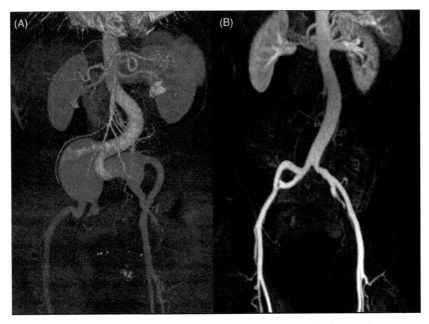

FIGURE 6C.2   Iliac artery aneurysm surgery. (A) A three-dimensional magnetic resonance angiography (MRA) shows bilateral large iliac artery aneurysms (69 and 42 mm). (B) A postoperative MRA shows relative stenoses of the distal anastomoses due to progressive tortuosity and elongation of the native iliac arteries. *(Reused with permission from the article by van der Linde et al. Aneurysm-osteoarthritis syndrome with visceral and iliac artery aneurysms. J Vasc Surg 2013;57:96–102.)*

The 35-year-old patient who underwent open repair of the splenic aneurysm (described earlier) presented 5 years after his surgery with abdominal pain [5]. CT angiography revealed a hepatic artery aneurysm that had grown from 11 to 18 mm within 1 year. Because this aneurysm had become symptomatic, it was excluded by implantation of a covered self-expandable stent graft (Viabahn; Gore, Flagstaff, Ariz) in the extrahepatic part of the hepatic artery, thereby closing off a second small saccular intrahepatic artery aneurysm (Fig. 6c.4) [5]. This procedure was uncomplicated, although follow-up CT angiography demonstrated a completely new fusiform aneurysm (15 × 11 mm) in the left gastric artery. For this issue, the patient successfully underwent coil embolization.

Other reports of endovascular repair of medium-size arteries in AOS patients are scarce. Chris Burke, Sherene Shalhub, and Benjamin Starnes reported on an endovascular repair of an internal mammary artery aneurysm in a patient with AOS [17]. This case report described a 49-year-old woman who presented with chest pain 2 years after undergoing a Bentall procedure for a type A dissection. She was found to have a large left mediastinal hematoma and a left internal mammary artery aneurysm (11 × 13 mm) without active extravasation on CT angiography. She underwent elective endovascular embolization with coils

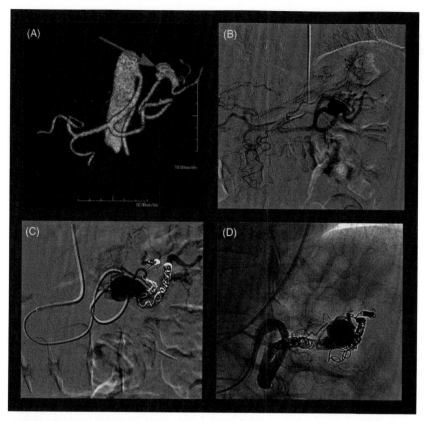

FIGURE 6C.3   Coil embolization of a splenic artery aneurysm. (A) A three-dimensional computed tomography angiography and (B) angiography showing a splenic artery aneurysm (21 mm) and tortuosity. Coil embolization procedure to occlude of the splenic artery (C) distal and (D) proximal to the aneurysm. *(Reused with permission from the article by van der Linde et al. Aneurysm-osteoarthritis syndrome with visceral and iliac artery aneurysms. J Vasc Surg 2013;57:96–102.)*

(Cook Medical, Bloomington, Ind.) [17]. There were no complications. Follow-up CT angiography demonstrated occlusion of the aneurysm.

6   RECOMMENDATIONS FOR OPEN AND ENDOVASCULAR INTERVENTIONS

Because AOS patients can experience dissections at relatively small diameters, early elective aneurysm repair seems to be appropriate to avoid vascular catastrophes. Nevertheless, potential benefits should always be weighed against the risks of a procedure. Before any type of intervention, one should aim to carefully evaluate aneurysmal growth, procedural complication rates, and interventional outcomes. Although aneurysm growth rate in AOS patients is still unknown, it has become clear that growth can be fast and unpredictable [5,15].

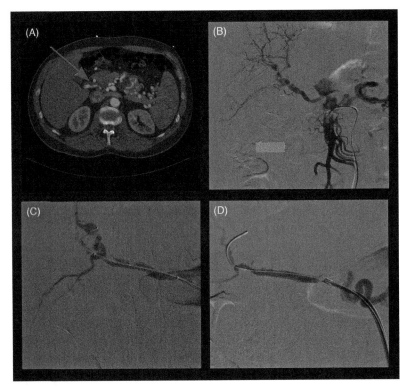

FIGURE 6C.4   Stenting of a hepatic artery aneurysm. (A) Computed tomography angiography and (B) angiography show a hepatic artery aneurysm (18 mm) and tortuosity. Angiography images showing different stages of the intervention: covered stent is (C) in place and (D) unfolded. *(Reused with permission from the article by van der Linde et al. Aneurysm-osteoarthritis syndrome with visceral and iliac artery aneurysms. J Vasc Surg 2013;57:96–102.)*

The current general consensus in atherosclerotic aneurysmal disease is that (endo)vascular treatment is indicated in asymptomatic visceral artery aneurysms > 2.0 cm and iliac artery aneurysms >3.0 cm [18–23]. However, due to the sometimes rapid aneurysmal growth and occurrence of dissections in only mildly dilated arteries, a more aggressive treatment strategy seems warranted in AOS patients [5,15].

So far, (endo)vascular treatment experience in AOS patients has been limited to seven described interventions due to the recent discovery of this syndrome [5,15,16]. In vascular-type EDS, friable vascular tissue leads to high surgical and endovascular complication rates [24]. In contrast, fragility of the arterial tissue seems not to be a major issue in the described surgical experience in AOS patients [5,17]. Tissue handling felt the same as in patients without a connective tissue disorder; thus, elective interventions seem to be feasible and safe in AOS patients [5,15].

Although endovascular treatment of aortic aneurysms is generally discouraged in patients with connective disorders, little is known about open versus endovascular repair of visceral aneurysms in patients with connective tissue disorders [25]. The potential harmful impact of persistent radial forces of a stent graft might be less of an issue in the visceral arteries than in the aorta and of no concern with coil embolization [5]. Furthermore, visceral aneurysms might be difficult to treat through an open surgical procedure, and periprocedural morbidity and mortality will generally be lower for endovascular procedures [5]. Therefore, the consensus is to go for an "endovascular first" strategy with AOS patients, although an individualized approach weighing all potential benefits and harms and a multidisciplinary evaluation before deciding on the treatment strategy are recommended [5]. Long-term follow-up is needed after repair and for surveillance for new aneurysmal formations [5].

7 CONCLUSIONS

Extensive imaging of the arterial tree identified (multiple) visceral and iliac artery aneurysms in 30–40% of AOS patients. Although surgical experience has been limited to case studies, fragility of the arterial tissue does not seem to complicate open and endovascular procedures for AOS patients. Due to the higher risk of dissection in mildly dilated arteries, a more aggressive treatment strategy seems to be appropriate. To prevent rupture, elective aneurysm repair should be considered in any visceral or iliac artery aneurysm that exceeds twice the expected arterial diameter or grows rapidly. After intervention, the entire arterial tree should be monitored frequently, because it remains at risk for aneurysm development and dissections or ruptures.

REFERENCES

[1] Stanley JC, Wakefield TW, Graham LM, Whitehouse WM Jr, Zelenock GB, Lindenauer SM. Clinical importance and management of splanchnic aneurysms. J Vasc Surg 1986;3(5):836–40.

[2] Carr SC, Pearce WH, Vogelzang RL, McCarthy WJ, Nemcek AA Jr, Yao JS. Current management of visceral artery aneurysms. Surgery 1996;120(4):627–33.

[3] Carr SC, Mahvi DM, Hoch JR, Archer CW, Turnipseed WD. Visceral artery aneurysm rupture. J Vasc Surg 2001;33(4):806–11.

[4] Wagner WH, Allins AD, Treiman RL, Cohen JL, Foran RF, Levin PM, et al. Ruptured visceral artery aneurysms. Ann Vasc Surg 1997;11(4):342–7.

[5] van der Linde D, Verhagen HJ, Moelker A, van de Laar IM, van Herzeele I, de Backer J, et al. Aneurysm-osteoarthritis syndrome with visceral and iliac artery aneurysms. J Vasc Surg 2013;57(1):96–102.

[6] Johnson PT, Chen JK, Loeys B, Dietz H, Fishman EK. Loeys-Dietz syndrome: MDCT angiography findings. Am J Roentgenol 2007;189(1):W29–35.

[7] Casey K, Zayed M, Greenberg JI, Dalman RL, Lee JT. Endovascular repair of bilateral iliac artery aneurysms in a patient with Loeys-Dietz syndrome. Ann Vasc Surg 2012;26(107):e5–10.

[8] Oderich G, Pannenton J, Bower T, Lindor NM, Cherry KJ, Noel AA, et al. The spectrum, management and clinical outcome of Ehlers-Danlos syndrome type IV: a 30-year experience. J Vasc Surg 2005;42(1):98–106.

[9] van de Laar IM, Oldenburg RA, Pals G, Roos-Hesselink JW, de Graaf BM, Verhagen JM, et al. Mutations in SMAD3 cause a syndromic form of aortic aneurysms and dissections with early-onset osteoarthritis. Nat Genet 2011;43(2):121–6.

[10] van der Linde D, van de Laar IM, Bertoli-Avella AM, Oldenburg RA, Bekkers JA, Mattace-Raso FU, et al. Aggressive cardiovascular phenotype of aneurysms-osteoarthritis syndrome caused by pathogenic SMAD3 variants. J Am Coll Cardiol 2012;60(5):397–403.

[11] van de Laar IM, van der Linde D, Oei EH, Bos PK, Bessems JH, Bierma-Zeinstra SM, et al. Phenotypic spectrum of the SMAD3-related aneurysms-osteoarthritis syndrome. J Med Genet 2012;49(1):47–57.

[12] Aubart M, Gobert D, Aubart-Cohen F, Detaint D, Hanna N, d'Indya H, et al. Early-onset osteoarthritis, Charcot-Marie-Tooth like neuropathy, autoimmune features, multiple arterial aneurysms and dissections: an unrecognized and life threatening condition. PLoS One 2014;9(5):e96387.

[13] van Hemelrijk C, Renard M, Loeys B. The Loeys-Dietz syndrome: an update for the clinician. Curr Opin Cardiol 2010;25(6):546–51.

[14] Loeys BL, Schwarze U, Holm T, Callewaert BL, Thomas GH, Pannu H, et al. Aneurysm syndromes caused by mutations in the TGF-beta receptor. N Engl J Med 2006;355(8):788–98.

[15] van der Linde D, Bekkers JA, Mattace-Raso FU, van de Laar IM, Moelker A, van den Bosch AE, et al. Progression rate and early surgical experience in the new aggressive aneurysms-osteoarthritis syndrome. Ann Thorac Surg 2013;95(2):563–9.

[16] Martens T, van Herzeele I, de Ryck F, Renard M, de Paepe A, François K, Vermassen F, De Backer J. Multiple aneurysms in a patient with aneurysms-osteoarthritis syndrome. Ann Thorac Surg 2013;95(1):332–5.

[17] Burke C, Shalhub S, Starnes BW. Endovascular repair of an internal mammary artery aneurysm in a patient with SMAD3 mutation. J Vasc Surg 2015;62(2):486–8.

[18] Van Petersen A, Meerwaldt R, Geelkerken R, Zeebregts C. Surgical options for the management of visceral artery aneurysms. J Cardiovasc Surg 2011;52(3):333–43.

[19] Pulli R, Dorigo W, Troisi N, Pratesi G, Innocenti AA, Pratesi C. Surgical treatment of visceral artery aneurysms: A 25-year experience. J Vasc Surg 2008;48(2):334–42.

[20] Abbas MA, Fowl RJ, Stone WM, Panneton JM, Oldenburg WA, Bower TC, et al. Hepatic artery aneurysm: factors that predict complications. J Vasc Surg 2003;38(1):41–5.

[21] Lakin RO, Bena JF, Sarac TP, Krajewski LP, Srivastava SD, Clair DG, et al. The contemporary management of splenic artery aneurysms. J Vasc Surg 2011;53(4):958–65.

[22] Santilli SM, Wernsing SE, Lee ES. Expansion rates and outcomes for iliac artery aneurysms. J Vasc Surg 2000;31(1 Pt. 1):114–21.

[23] Huang Y, Gloviczki P, Duncan AA, Kalra M, Hoskin TL, Oderich GS, et al. Common iliac artery aneurysm: expansion rate and results of open surgical and endovascular repair. J Vasc Surg 2008;47(6):1203–11.

[24] Oderich G, Pannenton J, Bower T, Lindor NM, Cherry KJ, Noel AA, et al. The spectrum, management and clinical outcome of Ehlers-Danlos syndrome type IV: a 30-year experience. J Vasc Surg 2005;42(1):98–106.

[25] Svensson LG, Kouchoukos NT, Miller DC, Bavaria JE, Coselli JS, Curi MA, et al. Expert consensus document on the treatment of descending thoracic aortic disease using endovascular stent-grafts. Ann Thorac Surg 2008;85(Suppl. 1):S1–41.

Chapter 6d

Orthopedic Evaluation and Treatment Options

P.K. Bos, MD, PhD

1 INTRODUCTION

Aneurysms-Osteoarthritis syndrome (AOS), a recently described autosomal dominant syndrome, is characterized by the presence of thoracic aneurysms and dissections and mild craniofacial, skeletal, and cutaneous abnormalities [1]. Musculoskeletal abnormalities associated with AOS include early-onset osteoarthritis, intervertebral disc degeneration, osteochondritis dissecans, mild craniofacial features, and meniscal anomalies. In this subchapter, the phenotypic presentation of patients with AOS, clinical evaluation and follow-up, and current orthopedic treatment options are discussed.

2 ETIOLOGY AND PATHOGENESIS

The human skeleton comprises bone, cartilage, joints, and ligaments. During embryology and development, trabecular and cortical bone is formed from embryonic mesenchymal tissue through intramembranous and endochondral ossification. In relation to degenerative bone and cartilage diseases, apart from other types of cartilage [eg, elastic cartilage (eg, ear) and fibrocartilage (eg, vertebral disc)], two distinct types of cartilage are of interest: growth plate cartilage and hyaline (articular) cartilage. During skeletal development, chondrocytes in growth plates undergo terminal differentiation: chondrocytes become hypertrophic and express type X collagen, MMP13, and osteocalcin; chondrocytes become apoptotic; and finally the matrix calcifies and is replaced by bone. In articular cartilage formation, terminal differentiation is blocked, which results in residual hyaline cartilage in the synovial joints. In osteoarthritis development, articular cartilage degenerates, showing characteristics similar to those seen in terminally differentiating chondrocytes in growth plates [2].

Intracellular SMAD3 protein acts as a downstream mediator in the transforming growth factor-beta (TGF-β) signaling pathway via membrane-bound type I and type II serine/threonine kinase receptors and their intracellular

Aneurysms-Osteoarthritis Syndrome. http://dx.doi.org/10.1016/B978-0-12-802708-0.00015-6
Copyright © 2017 Elsevier Inc. All rights reserved.

effectors, SMADs [2,3]. Chondrocyte terminal differentiation is stimulated by SMAD 1/5/8 activation and inhibited by the SMAD 2/3 pathway, where SMAD3 appears to have a more pronounced role [4]. Chondrocyte terminal differentiation via the SMAD pathways acts via Runx2. Runx2 is switched on or off by SMAD2/3 or SMAD1/5/8 [2].

Smad3 knockout mice develop degenerative joint disease resembling human osteoarthritis [5]. On the other hand, homozygous Runx2-deficient mice lack hypertrophic cartilage and have no bone formation [6]. Clinical and genetic research has shown that genetic variation in *SMAD3* is associated with hip and knee osteoarthritis relative to controls [7]. Furthermore, based on our recent description of early-onset osteoarthritis, mostly in multiple joints in patients with AOS, a large prospective population-based study of osteoarthritis and osteoporosis demonstrated a significant association of *SMAD3* defects with radiographic osteoarthritis [1,8]. These and future clinical data and knowledge from experimental research may advance development of new therapies targeting osteoarthritis and bone disease. Information about patient characteristics and the clinical presentation of AOS as well as management and treatment options is limited. Close follow-up by dedicated physicians will help further elucidate optimal treatment and care for these patients. Moreover, the study of rare bone diseases has been shown to lead to new mechanistic and therapeutic insights [9].

3 MUSCULOSKELETAL EVALUATION AND PHENOTYPIC PRESENTATION

All patients diagnosed with AOS in our clinic in Rotterdam are now evaluated by one dedicated pediatric or adult orthopedic surgeon. Each patient's medical history, previous treatment, and current musculoskeletal symptoms are taken into account. Physical examination is focused on signs of osteoarthritis, spinal disorders, osteochondritis dissecans, meniscal lesions, joint laxity, and craniofacial abnormalities. Routine radiological evaluation includes symptomatic sites, total spine, pelvis and hips, and knees. Magnetic resonance imaging (MRI) of the joints is performed in case of radiological abnormalities or unexplained symptoms. Following the patient's initial evaluation, the frequency of further follow-up is based on the severity of symptoms.

In the initially evaluated AOS patient group, most patients (96%) had radiologic signs of osteoarthritis, 75% in two or more joints [1]. In the majority of adult patients (19 of 35, or 54%), joint complaints (eg, osteochondritis dissecans, osteoarthritis and meniscal lesions) were the first symptom. This emphasizes the importance of awareness of cardiovascular abnormalities in patients with multiple joint complaints. In our clinic in Rotterdam, all patients with (familial) thoracic aortic aneurysms and dissections and musculoskeletal complaints are routinely screened by a dedicated orthopedic surgeon. Also, asymptomatic patients with thoracic aortic aneurysms and dissections, AOS patients, and family members not willing to be genetically screened are frequently referred for evaluation of musculoskeletal complaints. Radiologic evaluation is then guided by symptoms.

Orthopedic surgeons as well as cardiologists and vascular surgeons should be aware of these combinations of symptoms, especially in families with early death from aortic aneurysms. Similarly, we are accustomed to considering other systemic diseases with multiple joint symptoms, such as rheumatoid arthritis, hyperlaxity disorders (eg, Marfan, Ehlers-Danlos), gout, or systemic lupus erythematosus. The orthopedic surgeon responsible for the orthopedic care of these patients should be aware of the known skeletal features associated with AOS. Targeting the evaluation of current symptoms and informing patients and relatives about possible or future complaints are important.

Although the focus of treatment following the diagnosis of AOS in a patient or family is not based primarily on the musculoskeletal complaints, attention and awareness are needed for long-term care. Orthopedic knowledge is limited but growing, so it is important to gain experience by centralizing patient care. Most skeletal features when diagnosed are treated in ways that conform to current standard care. Life expectancy and possible multiple joint involvement may guide treatment choices.

4   OSTEOARTHRITIS

Joint complaints are often the first symptoms that cause patients to consult a physician. In none of the initially described patient cohort was an aneurysm connective tissue disease considered at first presentation [1]. This again emphasizes the need to inform physicians and increase their awareness of systemic connective tissue diseases, such as AOS.

Osteoarthritis is characterized by degeneration of the synovial joint tissues, including cartilage. It is a progressive disease, currently the most common form of arthritis, and a major cause of disability worldwide [10]. Its prevalence is expected to increase due to the ageing of Western society and with the increasing prevalence of obesity. Currently, there is no disease-modifying treatment for osteoarthritis. Symptomatic treatment consists of patient education, exercise therapy, pain medication, weight loss, and reduction of joint loading and injections [11–13]. Surgical treatment options for end-stage osteoarthritis include resection arthroplasty, arthrodesis, correction osteotomy to unload the affected joint compartment, and prosthetic joint replacement. The limited implant survival of joint arthroplasties, especially in young patients, results in prolonged conservative treatment. The life expectancy of AOS patients, with unaffected vascularity and thorough follow-up or following treatment for vascular anomalies or dissections, is probably not altered compared to the normal population. Therefore, we do not advise another approach for arthroplasty indications for AOS patients. Revision surgery due to loosening of implants, with the chance of challenging bone loss, is expected for young patients. Figs. 6d.1 and 6d.2 illustrate cases of osteoarthritis in AOS patients.

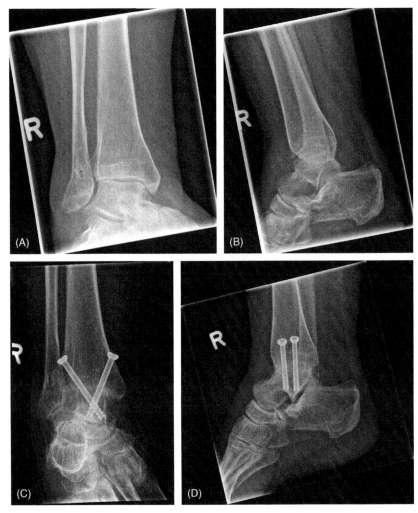

FIGURE 6D.1 41-year-old woman diagnosed with AOS and disabling ankle osteoarthritis (A&B) with a history of three lateral ankle stabilization procedures for ankle joint instability and arthroscopic losse body removal. She was treated with an arthroscopic ankle arthrodesis (C&D).

5 OSTEOCHONDRITIS DISSECANS

Osteochondritis dissecans is a condition of focal, idiopathic, subchondral bone lesions with possible involvement of the overlying cartilage. Osteochondritis dissecans affects joints of children and young adults and forms a common cause for loose bodies in the synovial joints, most often in the knees. Osteochondritis dissecans occurs in 15–29 people per 100,000 each year [14]. The etiology and pathogenesis of the disease remains largely unknown. Several proposed

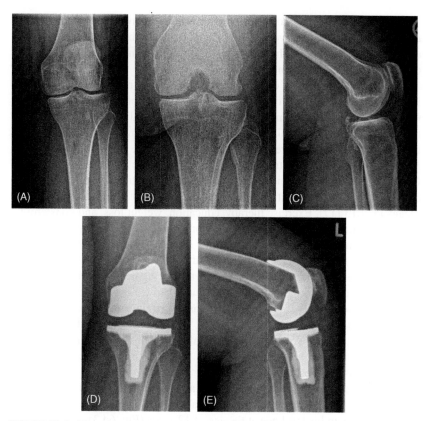

FIGURE 6D.2 The same woman, now 42 years old, with radiologic osteoarthritis, Kellgren Lawrence grade 2 (A–C: anterior–posterior weightbearing, Rosenberg, and lateral view). An arthroscopy 1 year earlier for locking complaints revealed chondopathy grades 3 and 4 in the medial compartment and grade 2 in the lateral and patellofemoral compartment. She was treated with a posterior stabilized total knee arthroplasty (D&E: anterior–posterior and lateral view).

causes for the disease include repetitive microtrauma to vulnerable parts of the joint surface, local ischemia leading to avascular necrosis, hereditary factors, and bone-formation abnormalities [15]. Recent animal genome-wide association studies have identified genes possibly involved in the disease process, including genes involved in extracellular matrix molecules and growth-plate maturation [15].

AOS is an example of a syndrome with a high prevalence of osteochondritis dissecans (> 50% of patients reported a history of osteochondritis dissecans in the initially described series) [1]. The high prevalence of osteochondritis dissecans and the genetic disorder, with its influence on bone and cartilage development, make AOS a clear example of genetic involvement in the osteochondritis dissecans disease process. Studying the features of this disease may help in further understanding osteochondritis dissecans and unrevealing its pathophysiology [9].

AOS may cause severe abnormalities in the *SMAD3* pathway and subsequent osteochondritis dissecans development, whereas milder abnormalities of *SMAD3* and the transforming growth factor-beta (TGF-β) signaling pathway may explain osteochondritis dissecans in other patients.

Clinical presentation of osteochondritis dissecans consists of effusion, activity-related knee pain with or without mechanical complaints, such as locking. Osteochondritis dissecans may also be an incidental radiological finding in patients with unrelated complaints. Knee, ankle, and elbow joints are most commonly involved. X-rays are used for initial evaluation, but often MRI is used for further evaluation and treatment planning [16]. The American Academy of Orthopaedic Surgeons defined recommendation for treatment of osteochondritis dissecans in its second-edition guidelines for treatment of osteochondritis dissecans in 2013 [17]. Exercise therapy, education, weight loss, and nonsteroidal antiinflammatory drugs are the most important recommendations for treating symptoms. There is a lack of evidence for surgical treatment options [16,17]. Nonoperative treatment should be evaluated within 3–6 months, because surgical debridement or fixation of bone and cartilage fragments may be warranted. The healing success of nonoperative treatment varies between 50% and –94% [16]. In general, surgery is considered for unstable fragments of following failed conservative treatment. Surgical treatment options include subchondral bone drilling, refixation of fragments, bone grafting, debridement, and alignment procedures to unload the affected joint area. Autologous chondrocyte implantation may be used to restore large defects.

As an example of treatment, we present a case of a 17-year-old AOS patient, who was conservatively treated for a medial femoral condyle knee osteochondritis dissecans in his left knee at the age of 12 years (Fig. 6d.3).

At the age of 17 years, he developed activity-related pain and effusion in his left knee. These complaints resolved with activity modification and nonsteroidal anti-inflammatory drugs. X-ray and MRI revealed an osteochondritis dissecans lesion at an unusual tibial plateau location (Fig. 6d.4). This was treated

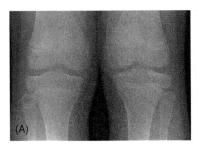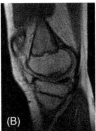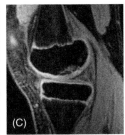

FIGURE 6D.3   12-year-old patient diagnosed with osteochondritis dissecans of the weight-bearing area of the medial femoral condyle. X-ray (A) and T1 and T2 magnetic resonance image (B and C, respectively) show a subchondral bone lesion with intact overlying articular cartilage. This patient was successfully treated conservatively.

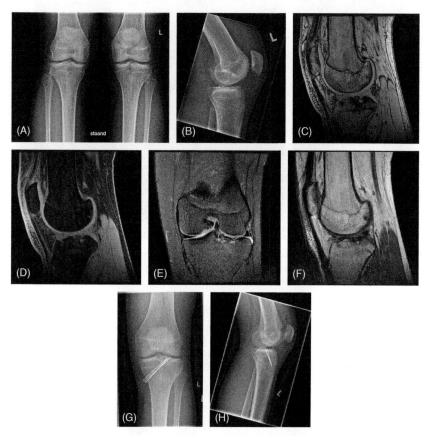

FIGURE 6D.4 Anteroposterior and lateral view showed subtle subchondral bone sclerosis in the medial tibial plateau of the left knee (A&B). Sagittal and coronal magnetic resonance images show subchondral bone changes, edema, and fluid between fragment and underlying bone (C–F). Postoperative anteroposterior and lateral view showing stabilization of the osteochondritis dissecans fragment with three temporary Kirschner wires (G&H).

with arthroscopy and stabilization of the fragment with Kirschner wires, which were removed after 6 weeks.

6 SPINAL DISORDERS

The majority of AOS patients have intervertebral disc degeneration, mainly involving the cervical and lumbar spine. Some vertebral body abnormalities, such as Schmorl's nodules (protrusion of the intervertebral disc through the vertebral endplate), can already be visualized in childhood (Fig. 6d.5). Spondylolysis and/or spondylolisthesis and facet or uncovertebral joint osteoarthritis are commonly found. Severe spondyloarthritis may lead to degenerative scoliosis (Fig. 6d.6). In our clinic in Rotterdam, we believe it is important to

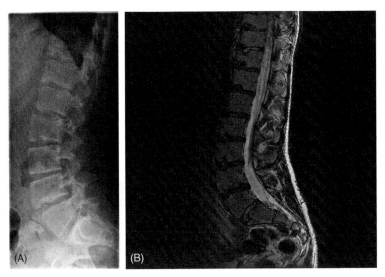

FIGURE 6D.5 18-year-old man with intervertebral disc degeneration and Schmorl's nodules at several levels (A). MRI scan showing endplate irregularities with protrusion of the intervertebral discus through the vertebral endplates (Schmorl's nodules) (B).

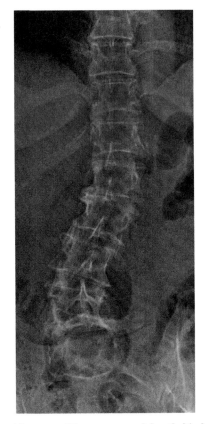

FIGURE 6D.6 53-year-old woman with a severe spondyloarthritis, leading to a degenerative scoliosis.

radiologically screen for spinal disorders with X-ray and MRI to reveal possible abnormalities at an early stage. Early education, exercise, and spinal load reduction may reduce the chance of early degenerative disease that would necessitate invasive treatment.

7 CONCLUSIONS

In conclusion, musculoskeletal abnormalities associated with AOS are common and require initial evaluation by dedicated orthopedic surgeons when such a diagnosis is considered. On the other hand, it is also advisable to evaluate musculoskeletal abnormal patients for vascular aneurysms, especially when vascular symptoms are present or there is a family history of aneurysms or dissections. Most skeletal features, when diagnosed, are treated in ways that conform to current standards of care. However, with growing experience in treating these patients, new insights may alter the future treatment of musculoskeletal complaints for patients with (familial) thoracic aortic aneurysms and dissections.

REFERENCES

[1] van de Laar IM, et al. Phenotypic spectrum of the SMAD3-related aneurysms-osteoarthritis syndrome. J Med Genet 2012;49(1):47–57.

[2] van der Kraan PM, et al. TGF-beta signaling in chondrocyte terminal differentiation and osteoarthritis: modulation and integration of signaling pathways through receptor-Smads. Osteoarthr Cartilage 2009;17(12):1539–45.

[3] Finnson KW, et al. TGF-b signaling in cartilage homeostasis and osteoarthritis. Front Biosci (Schol Ed) 2012;4:251–68.

[4] Hellingman CA, et al. Smad signaling determines chondrogenic differentiation of bone-marrow-derived mesenchymal stem cells: inhibition of Smad1/5/8P prevents terminal differentiation and calcification. Tissue Eng Pt A 2011;17(7–8):1157–67.

[5] Yang X, et al. TGF-beta/Smad3 signals repress chondrocyte hypertrophic differentiation and are required for maintaining articular cartilage. J Cell Biol 2001;153(1):35–46.

[6] Hecht J, et al. Detection of novel skeletogenesis target genes by comprehensive analysis of a Runx2(-/-) mouse model. Gene Expr Patterns 2007;7(1–2):102–12.

[7] Valdes AM, et al. Genetic variation in the SMAD3 gene is associated with hip and knee osteoarthritis. Arthritis Rheum 2010;62(8):2347–52.

[8] Aref-Eshghi E, et al. SMAD3 is associated with the total burden of radiographic osteoarthritis: the Chingford study. PLoS One 2014;9(5):pe97786.

[9] Tosi LL, Warman ML. Mechanistic and therapeutic insights gained from studying rare skeletal diseases. Bone 2015;76:67–75.

[10] Kotlarz H, et al. Osteoarthritis and absenteeism costs: evidence from US National Survey Data. J Occup Environ Med 2010;52(3):263–8.

[11] Hochberg MC, et al. American College of Rheumatology 2012 recommendations for the use of nonpharmacologic and pharmacologic therapies in osteoarthritis of the hand, hip, and knee. Arthrit Care Res 2012;64(4):465–74.

[12] McAlindon TE, et al. OARSI guidelines for the non-surgical management of knee osteoarthritis. Osteoarthr Cartilage 2014;22(3):363–88.

[13] Brown GA. AAOS clinical practice guideline: treatment of osteoarthritis of the knee: evidence-based guideline, 2nd edition. J Am Acad Orthop Sur 2013;21(9):577–9.

[14] Kramer DE, et al. Surgical management of osteochondritis dissecans lesions of the patella and trochlea in the pediatric and adolescent population. Am J Sport Med 2015;43(3):654–62.

[15] Bates JT, et al. Emerging genetic basis of osteochondritis dissecans. Clin Sport Med 2014;33(2):199–220.

[16] Edmonds EW, Polousky J. A review of knowledge in osteochondritis dissecans: 123 years of minimal evolution from König to the ROCK study group. Clin Orthop Relat R 2013;471(4):1118–26.

[17] Jevsevar DS. Treatment of osteoarthritis of the knee: evidence-based guideline, 2nd edition. J Am Acad Orthop Sur 2013;21(9):571–6.

Chapter 6e

Genetic Counseling

J.M.A. Verhagen, MD, I.M.B.H. van de Laar, MD, PhD

1 INTRODUCTION

Aneurysms-Osteoarthritis syndrome (AOS), or Loeys-Dietz syndrome type 3 (LDS3) [MIM 613795], is an autosomal-dominant condition with marked inter- and intrafamilial phenotypic variations caused by mutations in the *SMAD3* gene. Molecular genetic testing of *SMAD3* is important to obtain an accurate diagnosis of AOS, as symptoms can be mild or aspecific and overlap with those of other types of Loeys-Dietz syndrome (LDS1 [MIM 609192], LDS2 [MIM 610168], LDS4 [MIM 614816], LDS5 [MIM 615582]) and other heritable connective tissue disorders, such as Marfan syndrome [MIM 154700], vascular Ehlers-Danlos syndrome [MIM 130050], and arterial tortuosity syndrome [MIM 208050]. Furthermore, a molecular diagnosis provides the opportunity of carrier testing in asymptomatic family members and enables young couples to discuss their reproductive options. However, genetic testing can have potential negative ethical, legal, and social implications. During pretest genetic counseling, the potential risks, benefits, and limitations of genetic testing are discussed, facilitating autonomous decision making. In this chapter, we provide an overview of the key points that should be addressed during genetic counseling of families with known or suspected AOS.

2 COUNSELING PROCEDURE

In genetic counseling, an appropriately trained healthcare professional provides individuals and families with personalized genetic information to promote informed choices and adaptation to the genetic risk or condition [1]. The reasons for requesting genetic counseling for AOS broadly fall into two categories: (1) confirmation of a suspected diagnosis or differential diagnosis (see Chapter 4) in symptomatic patients, or (2) risk assessment for healthy family members (predictive testing) or future offspring. Pretest genetic counseling typically starts with clarifying the counselees' needs and expectations.

Aneurysms-Osteoarthritis Syndrome. http://dx.doi.org/10.1016/B978-0-12-802708-0.00016-8
Copyright © 2017 Elsevier Inc. All rights reserved.

2.1   Genetic Testing of Patients

To determine the likelihood of disease, a detailed personal and family history—including specific enquiry about arterial aneurysm and dissection, cardiac valve abnormalities, sudden cardiac death, joint complaints, and musculoskeletal abnormalities—and relevant medical records and autopsy reports should be obtained. The physician should perform physical examinations and recommend additional clinical investigations. Together, these clinical findings can raise the clinical suspicion of AOS. However, owing to AOS's significant phenotypic overlap with other heritable connective tissue disorders, it is often difficult to make a definite diagnosis solely on clinical grounds in an index patient, that is, the first member of a family who comes to the attention of a geneticist. In such a case, molecular genetic testing is required for diagnostic confirmation—yet the results of genetic tests are not always straightforward. As the mutation detection rate is less than 100%, a negative result in an affected person does not completely exclude the diagnosis of AOS. In some cases, molecular genetic testing results in the identification of genetic variants of unknown clinical significance (VUS). These variants cannot be used in medical management until their significance is established. Communicating these variants should be done with caution, as this information may lead to confusion and misinterpretation among patients and physicians [2]. Tests of the parents and/or other affected family members, shared data in public databases, and, if available, functional assays can help determine the pathogenicity of a VUS.

2.2   Genetic Testing of Family Members

Symptomatic family members can often be easily recognized by their clinical features, although confirmation using molecular genetic testing is highly recommended. In asymptomatic family members, on the other hand, molecular genetic testing can be used to rule out AOS. Prior to offering a molecular genetic test, a physician should discuss the natural history of AOS (including the age-dependent penetrance and variable expressivity, both between and within families), features of autosomal-dominant inheritance (ie, males and females are affected with equal frequency, both sexes can transmit the disorder, and each child of an affected parent has a 50% chance of inheriting the mutant allele), management and treatment strategies, and implications of molecular genetic testing. Nondirective counseling requires the provision of accurate, unbiased information and subsequent exploration of the counselees' perceptions and interpretations. Many authors, however, have questioned whether absolute nondirectiveness is desirable or even possible [3].

A molecular diagnosis can remove uncertainty and help a patient make informed medical and lifestyle decisions, such as those regarding medical check-ups and family planning (see Section 4). Furthermore, if a disease-causing mutation has been identified, other family members can be offered carrier testing to clarify their risks and prevent life-threatening complications. This is usually performed by stepwise identification and testing of family members known as "cascade screening." Although diagnosis of AOS provides the opportunity to determine the appropriate surveillance of a potential aggressive vascular disease, it is essential to recognize that carriers can have difficulty coping with the test results and sometimes experience feelings of depression, anxiety, or guilt. Disclosure of genetic information to family members may cause distress and influence family relationships. Test results can lead to stigmatization and discrimination within the community (eg, from insurance carriers or employers). Furthermore, uncertainty regarding the course of the disease in the individual patient can lead to feelings of insecurity.

Counselees should be encouraged to take their time to reach appropriate decisions. If an individual decides to pursue genetic testing, blood samples are drawn, and a posttest appointment is scheduled to discuss the test results. Following a genetic counseling session, a detailed written report summarizing the contents of the consultation is usually provided. Follow-up visits or psychosocial support should be offered if needed.

3   GENETIC TESTING IN MINORS

Genetic testing in minors requires special consideration, taking into account the limited decision-making capabilities of children and the potential lifelong implications of test results. Decisions regarding genetic testing should be made in the best interest of the child [4,5]. In the absence of direct medical benefit, genetic testing is usually postponed until adulthood to protect the child's future autonomy. In some instances, however, the benefits of testing in childhood may outweigh the harms. Genetic testing may be performed for diagnostic purposes in a child with physical signs or symptoms of AOS. Under these circumstances, genetic testing is similar to other medical diagnostic evaluations [6]. Genetic testing may also be requested for an asymptomatic child with a positive family history of AOS. Predictive genetic testing of minors is generally accepted for childhood-onset conditions if preventative or therapeutic measurements are available to reduce morbidity or mortality [7]. Accordingly, predictive testing seems justified in asymptomatic children at risk for AOS. Although clinical expression of AOS is highly variable and age-dependent, severe cardiovascular manifestations have been observed in early childhood [8,9]. Predictive genetic testing may therefore be offered from the first year of life. When testing is deferred or the underlying genetic defect in the family is still unknown, regular clinical follow-up by a multidisciplinary team can be used as an alternative approach. We strongly recommend that the testing of children always involve psychosocial care.

Parents (or legal guardians) have the authority to make decisions on behalf of their children. Nonetheless, children should be actively involved in the counseling and decision-making process, according to their capacities. In older children and adolescents, more emphasis should be put on their assent. However, as numerous individual variations exist in the level of maturity, the minor's competency to consent has to be judged on a case-by-case basis and should not be based solely on age. Learning a child's genetic status can remove parental uncertainty or anxiety. When testing is performed early in life, the genetic information can become part of an individual's identity. However, genetic testing may have positive or negative effects on family relationships. Upon finding out that their child is a mutation carrier, some parents become more anxious and overprotective. Children identified as mutation carriers of inherited cardiovascular diseases generally display effective coping behavior. Nevertheless, dealing with the impact of the condition on their daily life may remain difficult, warranting the continued availability of psychosocial support [10].

4   REPRODUCTIVE OPTIONS

Young adults with AOS are faced with a number of issues and challenges, such as employment, social participation, and family planning. Timely referral for genetic counseling—preferably before pregnancy—is indicated so that prospective parents who are affected or at risk can be fully informed about the maternal and fetal complications of pregnancy, the recurrence risk for offspring, and their reproductive options. So far, however, data regarding reproductive decision making and pregnancy outcomes in patients with AOS are scarce. Further investigation in this direction is clearly needed to optimize preconception counseling. Women with AOS are considered at high risk of aortic complications during pregnancy and in the early postpartum period. Aortic dissection may occur at relatively small aortic diameters [11]. Prophylactic surgery before pregnancy is recommended when the ascending aorta exceeds 45 mm [12]. Women with aortic pathology should be monitored carefully by an expert cardiologist and obstetrician. Blood pressure control is of primary importance. Vaginal delivery is preferred if the aortic diameter is less than 40 mm. In women with an aortic diameter exceeding 45 mm, caesarean section is advised [13].

Several reproductive options are available for at-risk couples: (1) natural conception, (2) prenatal diagnosis (PND), (3) preimplantation genetic diagnosis (PGD), (4) gamete (egg or sperm) donation, (5) adoption, or (6) refraining from having (more) children. Many couples decide to accept the recurrence risk and prefer to conceive naturally, taking into account the wide phenotypic variation, the lack of clear genotype–phenotype correlation, and/or their personal moral, ethical, or religious values. In general, couples who experience more severe disease manifestations are more likely to request a PND or a PGD, which requires prior identification of the disease-causing mutation in the family. Prenatal ultrasound examination allows early detection of fetal abnormalities suggestive of AOS, such as aortic root dilatation. However, the sensitivity is very low, that is, the disease cannot be ruled out in the absence of sonographic markers, despite detailed evaluation.

PND involves an invasive procedure, such as chorionic villus sampling (CVS) or amniocentesis, to determine whether a fetus is affected with a monogenic or chromosomal disorder prior to birth. CVS usually takes place at 10–12 weeks of gestation. The sampling procedure can be performed using either a transabdominal or transcervical approach and mainly depends on the physician's experience and the position of the placenta. Amniocentesis is routinely performed after 15 weeks of gestation. Both procedures are associated with low risk of serious complications. The procedure-related risk of miscarriage following CVS is slightly higher compared to amniocentesis [14]. When prenatal testing reveals that the unborn child is affected, parents may opt for termination of pregnancy (TOP) or use the time to prepare themselves for a child with special needs. Requests for PND for heritable connective tissue disorders are uncommon, and controversy exists among patients and health care professionals regarding its use for the purpose of TOP [15–17]. Furthermore, as each child of an individual with AOS has a 50% chance of inheriting the *SMAD3* mutation, repeated pregnancy terminations may be necessary because of consecutively affected offspring.

PGD involves the use of in vitro fertilization (IVF) and the genetic testing of embryos prior to implantation. Only apparently unaffected embryos are selected for transfer into the uterus. However, several technical difficulties, such as the polymerase chain reaction–based DNA amplification from single cells, may compromise diagnostic accuracy. Couples are therefore advised to undergo PND because of the low risk of misdiagnosis. In general, PGD is considered more acceptable and advantageous, as it prevents transmission to future generations and avoids the potential need for pregnancy termination. On the other hand, PGD is a stressful and time-consuming procedure, and chances of success are relatively low, with a pregnancy rate of 22% per oocyte retrieval and 29% per embryo transfer for monogenic disorders [18]. No data have been published on possible adverse health effects of ovarian stimulation for IVF in women with AOS or other heritable connective tissue disorders.

REFERENCES

[1] Resta R, Biesecker BB, Bennett RL, Blum S, Hahn SE. National Society of Genetic Counselors' Definition Task Force. A new definition of genetic counseling: National Society of Genetic Counselors' Task Force report. J Genet Couns 2006;15(2):77–83.

[2] Soden SE, Farrow EG, Saunders CJ, Lantos JD. Genomic medicine: evolving science, evolving ethics. Pers Med 2012;9(5):523–8.

[3] Weil J, Ormond K, Peters J, Peters K, Biesecker BB, LeRoy B. The relationship of nondirectiveness to genetic counseling: report of a workshop at the 2003 NSGC Annual Education Conference. J Genet Couns 2006;15(2):85–93.

[4] European Society of Human Genetics. Genetic testing in asymptomatic minors: recommendations of the European Society of Human Genetics. Eur J Hum Genet 2009;17(6):720–1.

[5] American Academy of Pediatrics, Committee on Bioethics, Committee on Genetics, and American College of Medical Genetics and Genomics Social, Ethical and Legal Issues

Committee. Ethical and policy issues in genetic testing and screening of children. Pediatrics 2013;131(3):620–2.

[6] Ross LF, Saal HM, David KL, Anderson RR. American Academy of Pediatrics; American College of Medical Genetics and Genomics. Technical report: ethical and policy issues in genetic testing and screening of children. Genet Med 2013;15(3):234–45.

[7] Borry P, Evers-Kiebooms G, Cornel MC, Clarke A, Dierickx K. Public and Professional Policy Committee (PPPC) of the European Society of Human Genetics (ESHG). Genetic testing in asymptomatic minors: background considerations towards ESHG Recommendations. Eur J Hum Genet 2009;17(6):711–9.

[8] Fitzgerald KK, Bhat AM, Conard K, Hyland J, Pizarro C. Novel SMAD3 mutation in a patient with hypoplastic left heart syndrome with significant aortic aneurysm. Case Rep Genet 2014;2014:591516.

[9] Wischmeijer A, van Laer L, Tortora G, Bolar NA, van Camp G, Fransen E, Peeters N, di Bartolomeo R, Pacini D, Gargiulo G, Turci S, Bonvicini M, Mariucci E, Lovato L, Brusori S, Ritelli M, Colombi M, Garavelli L, Seri M, Loeys BL. Thoracic aortic aneurysm in infancy in aneurysms-osteoarthritis syndrome due to a novel SMAD3 mutation: further delineation of the phenotype. Am J Med Genet A 2013;161A(5):1028–35.

[10] Meulenkamp TM, Tibben A, Mollema ED, van Langen IM, Wiegman A, de Wert GM, de Beaufort ID, Wilde AA, Smets EM. Predictive genetic testing for cardiovascular diseases: impact on carrier children. Am J Med Genet A 2008;146A(24):3136–46.

[11] van der Linde D, van de Laar IM, Bertoli-Avella AM, Oldenburg RA, Bekkers JA, Mattace-Raso FU, van de Meiracker AH, Moelker A, van Kooten F, Frohn-Mulder IM, Timmermans J, Moltzer E, Cobben JM, van Laer L, Loeys B, De Backer J, Coucke PJ, De Paepe A, Hilhorst-Hofstee Y, Wessesl MW, Roos-Hesselink JW. Aggressive cardiovascular phenotype of aneurysms-osteoarthritis syndrome caused by pathogenic SMAD3 variants. J Am Coll Cardiol 2012;60(5):397–403.

[12] Regitz-Zagrosek V, Blomstrom Lundqvist C, Borghi C, Cifkova R, Ferreira R, Foidart JM, Gibbs JS, Gohlke-Baerwolf C, Gorenek B, Iung B, Kirby M, Maas AH, Morais J, Nihoyannopoulos P, Pieper PG, Presbitero P, Roos-Hesselink JW, Schaufelberger M, Seeland U, Torracca L. ESC Committee for Practice Guidelines. ESC Guidelines on the management of cardiovascular diseases during pregnancy: the Task Force on the Management of Cardiovascular Diseases during Pregnancy of the European Society of Cardiology (ESC). Eur Heart J 2001;32(24):3147–97.

[13] van Hagen IM, Roos-Hesselink JW. Aorta pathology and pregnancy. Best Pract Res Cl Ob 2014;28(4):537–50.

[14] Akolekar R, Beta J, Picciarelli G, Ogilvie C, D'Antonio F. Procedure-related risk of miscarriage following amniocentesis and chorionic villus sampling: a systematic review and meta-analysis. Ultrasound Obst Gyn 2015;45(1):16–26.

[15] Peters KF, Kong F, Hanslo M, Biesecker BB. Living with Marfan syndrome III. Quality of life and reproductive planning. Clin Genet 2002;62(2):110–20.

[16] Loeys B, Nuytinck L, Van Acker P, Walraedt S, Bonduelle M, Sermon K, Hamel B, Sanchez A, Messiaen L, De Paepe A. Strategies for prenatal and preimplantation genetic diagnosis in Marfan syndrome (MFS). Prenatal Diag 2012;22(1):22–8.

[17] Coron F, Rousseau T, Jondeau G, Gautier E, Binquet C, Gouya L, Cusin V, Odent S, Dulac Y, Plauchu H, Collignon P, Delrue MA, Leheup B, Joly L, Huet F, Theyenon J, Mace G, Cassini C, Thauvin-Robinet C, Wolf JE, Hanna N, Sagot P, Boileau C, Faivre L. What do French patients and geneticists think about prenatal and preimplantation diagnosis in Marfan syndrome? Prenatal Diag 2012;32(13):1318–23.

[18] Harper JC, Wilton L, Traeger-Synodinos J, Goossens V, Moutou C, SenGupta SB, Pehlivan Budak T, Renwick P, De Rycke M, Geraedts JP, Harton G. The ESHRE PGD Consortium: 10 years of data collection. Hum Reprod Update 2012;18(3):234–47.

Chapter 6f

Approach to Clinical Management

D. van der Linde, MD, MSc, PhD, J.W. Roos-Hesselink, MD, PhD

1 INTRODUCTION

Although Aneurysms-Osteoarthritis syndrome (AOS) has only been recently discovered and its the full spectrum and progression need to be clarified, some preliminary suggestions may be derived from the literature thus far [1–8]. Because multisystem involvement is frequently observed, cooperation in a multidisciplinary team with clinical geneticists, cardiologists, orthopedic surgeons, (interventional) radiologists, neurologists, vascular internists, and, when necessary, (vascular or cardiothoracic) surgeons is important. Because AOS was discovered at Erasmus University Medical Center in Rotterdam, the Netherlands, a multidisciplinary team here has proposed an algorithm for clinical surveillance [6].

2 INITIAL SCREENING

Cardiologists should suspect AOS in every patient with familial thoracic aortic aneurysms and dissections without molecular diagnosis or known cause, and such a patient should be tested for a genetic cause, including *SMAD3* mutations. For genetic screening and counseling options, please refer to Chapters 1 and 6e of this book. Furthermore, we suggest that clinicians treating patients with arterial aneurysmal disease in any large artery (eg, intracranial, iliac, splenic artery) at least ask whether these patients report joint symptoms. General practitioners should be aware of patients and families with a high rate of aortic dissections and sudden death at a young age in combination with joint complaints [6].

Upon the first physical examination of a patient suspected of familial thoracic aortic aneurysms and dissections, special attention must be paid to the presence of AOS-related findings, such as joint anomalies and craniofacial abnormalities, such as an abnormal uvula of hypertelorism [6]. A family history of dissections and sudden death at a young age is important. If patients report joint complaints, they should be referred to an orthopedic surgeon for diagnostic tests and further treatment (see Chapter 6d).

Aneurysms-Osteoarthritis Syndrome. http://dx.doi.org/10.1016/B978-0-12-802708-0.00017-X
Copyright © 2017 Elsevier Inc. All rights reserved.

As with every cardiac patient, an electrocardiogram should be performed, and the patient's blood pressure should be measured. Smoking should be discouraged. Stringent control of hypertension to limit aortic wall stress is recommended (see Chapter 6a) [2]. Furthermore, hypercholesterolemia should be treated aggressively.

All patients with AOS require transthoracic echocardiography to evaluate the diameters of the aortic root, left ventricular function, left ventricular hypertrophy, valve abnormalities, and congenital heart defects. To identify arterial aneurysms and tortuosity throughout the arterial tree, extensive cardiovascular baseline evaluation using computed tomography or magnetic resonance imaging from head to pelvis is recommended in every adult AOS patient [2]. If 50% risk carriers do not want presymptomatic gene testing, we suggest the same initial cardiovascular screening as for known mutation carriers [2,3].

3   MONITORING AND REFERRAL

Initially, full vascular imaging is recommended 1 year after baseline to establish the rate of progression or the formation of new aneurysms [2]. Thereafter, the imaging frequency should be guided by the findings—for instance, it should be done annually if the aortic diameter is more than 35 mm, if the aortic diameter shows significant growth (> 5 mm/year), or if there are large or growing aneurysms in other arteries throughout the arterial tree [2]. If no significant aneurysms are found, full vascular imaging are recommended at a 2-year intervals due to the unpredictable nature of the syndrome [2,6]. Transthoracic echocardiography is only repeated upon indication, depending on the results of the first echocardiogram [2].

Because dissections in AOS patients can occur at relatively small aortic diameters, early elective surgical intervention is recommended to reduce the risk of mortality. Patients should be referred to a cardiothoracic surgeon if the aortic root approaches 42 mm or is rapidly expanding (>5 mm/year) for evaluation of need for aortic root replacement (Chapter 6b) [2,4]. For aneurysms in other small- to medium-size arteries throughout the body, individual size or rate of growth and location must determine the treatment strategy (see Chapter 6c) [2,5]. Patients should be referred to an experienced vascular surgical team in collaboration with interventional radiologists.

Currently, the risk of rupture of intracranial aneurysms associated with AOS is unknown [1–3,7]. No deaths resulting from intracranial hemorrhage occurred in the Rotterdam cohort, but they were reported in the cohort by Regalado et al. [1–3,7]. If any intracranial or vertebral aneurysms are found, patients should be referred to a specialized neurologist for evaluation and treatment. Size, location, and rate of growth of the intracranial aneurysm are the most important determinants to decide whether intervention is needed [2].

At Erasmus University Medical Center in Rotterdam, we have implemented a clinical surveillance protocol for AOS patients, providing general guidelines about follow-up frequency and when consultation of other specialists and/or surgery is required (Fig. 6f.1) [6].

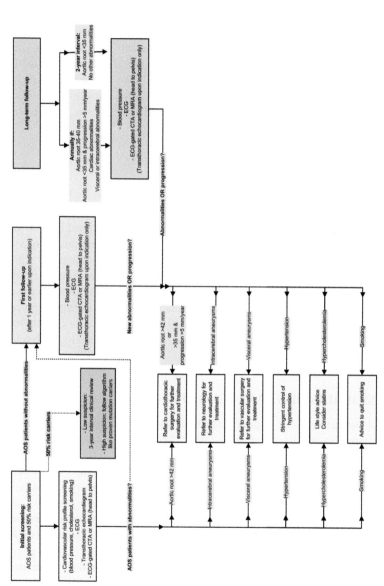

FIGURE 6F.1 Algorithm for clinical follow-up recommendations based on the multidisciplinary team expert consensus at Erasmus University Medical Center in Rotterdam in the Netherlands.

4 CHILDREN

The phenotype seems to be age-dependent, with aneurysms mainly and dissections only occurring in adulthood; however, the Rotterdam cohort did include six children [1–3]. Concerning screening and monitoring in childhood, clear suggestions are difficult to formulate at this time. It is recommended that all children should be referred to a specialized pediatric cardiologist. For children, we suggest transthoracic echocardiography for the initial screening around the age of 7 and magnetic resonance upon indication. We suggest that frequency of cardiologic evaluation with transthoracic echocardiography, magnetic resonance imaging, or both must be guided by the aortic root z-score and the presence of other cardiac abnormalities [2].

5 PSYCHOSOCIAL ADJUSTMENT

A genetic diagnosis, such as AOS, is associated with a significant health burden and an uncertain prognosis. Therefore, it can result in a variety of emotional reactions in patients and their loved ones. It can affect relationships between family members. Furthermore, the diagnosis of an autosomal-dominant syndrome has an important impact on family planning and the relationship with a spouse. For young and active patients in the prime of their life, exercise restrictions, such as avoidance of competitive contact sports or isometric exercises, may have a significant impact on their life choices and occupational choices. Physicians should be aware of signs of depression, isolated behavior, or anxiety. Attention must be paid to how patients and their families are coping, with referrals to counselors offered as necessary.

6 A MULTIDISCIPLINARY TEAM APPROACH

An important goal for the future is to provide patient-centered care, addressing AOS patients in an integral manner rather than focusing on specific organs. AOS commonly affects several organ systems, and therefore a multidisciplinary approach is applicable. This multidisciplinary team should consist of dedicated clinical geneticists, cardiologists, orthopedic surgeons, (interventional) radiologists, neurologists, vascular and cardiothoracic surgeons, and internists. Consultation between members of this team should be easy and fast, preferably within one academic medical center, to avoid keeping patients waiting in uncertainty.

Over the past year, some important steps in this direction have been made in the Erasmus Medical Center Rotterdam in the Netherlands with the initiation of an outpatient clinic specializing in genetic aortic disease. One specialized nurse practitioner (under supervision of cardiologists) is the central, easily approachable contact for the patients. This nurse practitioner also pays attention to the psychosocial well-being of the patients and regulates the multidisciplinary consults. If necessary, she can organize more intensive psychological support.

From our personal experience, we have noticed that general practitioners and emergency doctors are not completely aware of the high risk of complications in AOS patients when patients present with acute complaints. To ensure safe emergency care, we give all AOS patients an "SOS-letter" that briefly explains this rare syndrome to doctors who are not familiar with AOS, informs and warns them of the associated risks of vascular complications, and contains personal information about the known AOS-related abnormalities of that specific patient.

7 CONCLUSIONS

Because AOS affects multiple organ systems in various ways, patient care should be provided in a multidisciplinary team centered around each individual patient. In view of the aggressive nature of the vascular abnormalities, imaging of the entire arterial tree and transthoracic echocardiography is recommended for each patient and should be repeated at 1- or 2-year intervals, depending on the findings. Early referral to vascular surgeons, cardiothoracic surgeons, and neurologists in case of abnormalities is recommended. Furthermore, special attention must be paid to the psychosocial adjustment of patients and their families, with referrals to counselors as appropriate. These recommendations for the clinical surveillance of patients with AOS are based on the consensus of the multidisciplinary team at Erasmus Medical Center Rotterdam in the Netherlands.

REFERENCES

[1] van de Laar IM, Oldenburg RA, Pals G, Roos-Hesselink JW, de Graaf BM, Verhagen JM, et al. Mutations in SMAD3 cause a syndromic form of aortic aneurysms and dissections with early-onset osteoarthritis. Nat Genet 2011;43(2):121–6.

[2] van der Linde D, van de Laar IM, Bertoli-Avella AM, Oldenburg RA, Bekkers JA, Mattace-Raso FU, et al. Aggressive cardiovascular phenotype of aneurysms-osteoarthritis syndrome caused by pathogenic SMAD3 variants. J Am Coll Cardiol 2012;60(5):397–403.

[3] van de Laar IM, van der Linde D, Oei EH, Bos PK, Bessems JH, Bierma-Zeinstra SM, et al. Phenotypic spectrum of the SMAD3-related aneurysms-osteoarthritis syndrome. J Med Genet 2012;49(1):47–57.

[4] van der Linde D, Bekkers JA, Mattace-Raso FU, van de Laar IM, Moelker A, van den Bosch AE, et al. Progression rate and early surgical experience in the new aggressive aneurysms-osteoarthritis syndrome. Ann Thorac Surg 2013;95(2):563–9.

[5] van der Linde D, Verhagen HJ, Moelker A, van de Laar IM, van Herzeele I, de Backer J, et al. Aneurysm-osteoarthritis syndrome with visceral and iliac artery aneurysms. J Vasc Surg 2013;57(1):96–102.

[6] van der Linde D, van de Laar I, Moelker A, Wessels MW, Bertoli-Avella AM, Roos-Hesselink JW. Patients with aneurysms and osteoarthritis: Marfan syndrome ruled out, so what is it? Ned Tijdschr Geneeskd 2013;157(21):A5588.

[7] Regalado ES, Guo DC, Villamizar C, Avidan N, Gilchrist D, McGillivray B, et al. Exome sequencing identifies SMAD3 mutations as a cause of familial thoracic aortic aneurysm and dissection with intracranial and other arterial aneurysms. Circ Res 2011;109(6):680–6.

[8] Aubart M, Gobert D, Aubart-Cohen F, Detaint D, Hanna N, d'Indya H, et al. Early-onset osteoarthritis, Charcot-Marie-Tooth like neuropathy, autoimmune features, multiple arterial aneurysms and dissections: an unrecognized and life threatening condition. PLoS One 2014;9(5):e96387.

Index

Printed and bound by CPI Group (UK) Ltd, Croydon, CR0 4YY

08/05/2025

01865000-0001